Introduction to Statistics for Process Studies

Second Edition

Kim I. Melton

North Georgia College & State University

McGraw-Hill, Inc.

College Custom Series

New York St. Louis San Francisco Auckland Bogotá
Caracas Lisbon London Madrid Mexico Milan Montreal
New Delhi Paris San Juan Singapore Sydney Tokyo Toronto

McGraw-Hill's College Custom Series consists of products that are produced from camera-ready copy. Peer review, class testing, and accuracy are primarily the responsibility of the author(s).

INTRODUCTION TO STATISTICS FOR PROCESS STUDIES

ISBN 0-07-041449-1

Editor: J. D. Ice

21 22 23 24 25 26 27 28 29 30 GDP GDP 0 9 8 7 6 5 4 3 2
Printer/Binder Greyden Press

Preface

Business leaders are recognizing that quality has become an expectation from customers. Quality cannot be delegated to a small group within the company–all employees must be involved in quality improvement efforts. This means that process improvement concepts have made their way into offices at every level of organizations (from factory floor to board room). Statistics play a key role in understanding and improving processes. But, process studies have been omitted from traditional business statistics courses and textbooks.

This book was developed for students in BUS 301 (Business Statistics 1) at Virginia Commonwealth University. For two years a group of faculty from the Decision Sciences and Business Law Department worked to develop and test a new syllabus for this course. Conversations with faculty from other universities and with industry leaders have emphasized the need for a course that helps students develop skills in formulating questions, and collecting, analyzing, and interpreting data for making decisions in a business environment (many of which involve process studies).

This shift in thinking is causing evolution and revolution in the business statistics community. Textbook authors are scrambling to provide coverage, publishers are attempting to insure proper review of new material, and schools are revising curricula. Comprehensive new books with adequate coverage are not on the market (yet). Rather than wait and follow the lead of others, these materials have been created to supplement the current textbook. Familiarity with the topics covered in this book should provide students with a competitive advantage in the job market.

The first edition of this book, used during the 1991-92 school year, was well received. This second edition attempts to clarify some wording and present the material in a more readable format. The use of examples geared toward college juniors has been maintained. No references are given to the specific textbook that this supplements. In fact, references to "the textbook" are so general that most, if not all, current business statistics books will cover the topics mentioned.

Acknowledgments

Writing a book requires assistance from a number of people. A few people deserve special recognition. Without their help, the product that you see would be quite different.

- Cynthia C. Kozak, a 1991 graduate from Virginia Commonwealth University, for her input into the content of the book and for providing feedback from a student's viewpoint

- Dr. Charles H. Smith, Associate Professor at Virginia Commonwealth University, and Mary Graybeal, doctoral student at Virginia Commonwealth University, for their help in proofreading and wording
- Dot Haufler and Jill Kramer, Virginia Commonwealth University, for their recommendations on layout
- Dr. Robert L. Andrews, Associate Professor at Virginia Commonwealth University, for his careful review and feedback of material contained in the first edition
- All the other people who have read and commented on earlier versions of this book

Presentation of material assumes familiarity with the topics. Without the guidance of excellent teachers, this book would not have been possible. I am pleased to have had the opportunity to learn from some great teachers. Dr. W. Edwards Deming and Dr. Gipsie B. Ranney have (forever) changed the way I view statistics. Hopefully, this book will help you recognize what they have shown me: that statistics can be exciting and useful.

Kim I. Melton
June 4, 1992

Table of Contents

You need not be a mathematical statistician to do good statistical work . . . The statistical method is only good science brought up to date by the recognition that all laws are subject to the variations which occur in nature. Your study of statistical methods will not displace any other knowledge that you have; rather, it will extend your knowledge of engineering, chemistry, or economics, and make it more useful.

W. Edwards Deming

Introduction

In the last decade the role of statistics in business has been evolving. Until recently many people were introduced to the field of statistics through one or more courses, but the use of statistics was usually delegated to a small group of specialists within organizations. This group was often viewed with skepticism. They collected data, put it into a computer, pushed a few keys, drew conclusions, and reported back to the person or group that requested the study.

Today, statistics and statistical concepts are being used by people in all levels and positions within organizations. This shift has resulted, at least in part, from the need for companies to remain competitive in a world market. Managers have recognized that companies consistently providing superior products and services gain customer loyalty, increase profits, and ensure employment for their workers. The concept of consistency and the aim of statistics–to understand variation–go hand in hand.

How does this change affect the business statistics courses taught at universities? In the past, most business statistics courses introduced students to a collection of techniques. These techniques could be used to summarize data or to draw conclusions about a characteristic of some collection of objects by studying this characteristic in a "small" portion, or sample, of these objects. These areas of statistics have been labelled Descriptive Statistics and Inferential Statistics. Increasingly, textbooks have provided more "real world" data. Omitted from most current books is an introduction to collecting data. Also omitted is an introduction to the tools and techniques that are appropriate for studying processes and planning for the future output from those processes.

This supplement to your textbook will introduce these topics. The material will be presented in modules. Each module will include a discussion of the important concepts, "real world" examples, and exercises that allow you the opportunity to apply the concepts. The modules are:

Mod 1: Purpose of Statistical Studies: Describe, Infer, Predict

Mod 2: Statistical Thinking

Mod 3: Introduction to Process Studies

Mod 4: Process Flow Diagrams

Mod 5: Cause and Effect Diagrams

Mod 6: Check Sheets

Mod 7: Pareto Charts

Mod 8: Run Charts

Mod 9: Control Charts

Mod 10: Histograms

Mod 11: Scatter Diagrams

Mod 12: Data Collection

Successful use of the tools and concepts presented here assumes mathematical knowledge at the elementary algebra level and no previous exposure to statistics. Graphical displays will be used to clarify and explain concepts.

In the space of one hundred and seventy-six years the Lower Mississippi has shortened itself two hundred and forty-two miles. This is an average of a trifle over one mile and a third per year. Therefore, any calm person, who is not blind or idiotic, can see that in the Old Oolitic Silurian Period, just a million years ago next November, The Lower Mississippi River was upward of one million three hundred thousand miles long, and stuck out over the Gulf of Mexico like a fishing rod. And by the same token any person can see that seven hundred and forty-two years from now the Lower Mississippi will be only a mile and three-quarters long, and Cairo and New Orleans will have joined their streets together . . . There is something fascinating about science. One gets such wholesale returns of conjecture out of such a trifling investment of fact.

Mark Twain

Module 1: Purpose of Statistical Studies

Most people enter their first course in business statistics with very little idea what to expect. They are not sure what statistic is (or are), they question their math background, and they think that probability and statistics are interchangeable words. Their previous exposure to the field of statistics consists of hearing the statistician listed in the credits for professional football games (and other sports) and being bombarded with survey results that quote various "statistics." These represent two of the multitude of applications for statistics. Many more exist.

What is Statistics?

Simply stated, statistics is a field of study that deals with the study of variability. The study of variability includes recognizing when and where variability can be seen, describing variability that is present in some phenomenon, identifying sources of variability, determining how to minimize the effects of various sources of variability, determining what happens as a result of variability, and using the knowledge gained as a basis for rational decision making. The collection and analysis of data is inherent in accomplishing these tasks.

Historically, business statistics courses have focused on two types of statistical studies–**Descriptive Studies** and **Inferential Studies**. A third type of statistical study also deserves attention. This type of study is referred to as an **Analytic Study**. Statistics can be used to 1) describe, 2) infer, or 3) predict. A general description of each of these uses will be provided. Your textbook will provide more in-depth study of the first two types of studies. The additional modules included here will focus on the third with an emphasis on using process studies to predict future outcomes.

Descriptive Statistics

The goal of descriptive statistics is to summarize information about some characteristic(s) of a collection of objects. These objects could be people, papers, figures, events, etc. Descriptive statistics do **not** attempt to generalize the results shown (or reported) to any other collection of objects. Summary statistics may take the form of numbers or graphs. The following example will illustrate some important concepts in the use of descriptive statistics. Additional discussion and examples can be found in your textbook.

Example 1: A teacher has two sections of the same class that have taken a test. The tests have been graded and recorded as shown in Figure 1.1.

Grades on Test 1

Section 1		Section 2	
54	100	54	76
54	100	70	78
54	100	72	84
54	100	73	89
54	100	74	100
Average = 77		Average = 77	

Figure 1.1 Grades Listed by Section

As the teacher enters the room, a student asks, "How were the tests?" Consider some possible responses that the teacher could provide; the appropriateness of each approach will be discussed after viewing the alternatives. First, the teacher could summarize the results with the following statement. "The grades ranged from 54 to 100 with an average of 77." Second, the teacher could summarize each section's results graphically as shown in Figure 1.2. Or finally, the teacher could reason that these were two sections of the same course, that the same lecture had been used in both sections, and that the same expectations apply to both groups of students. This logic leads to a conclusion to combine the results from the two sections into one graph. The resulting graph is shown in Figure 1.3.

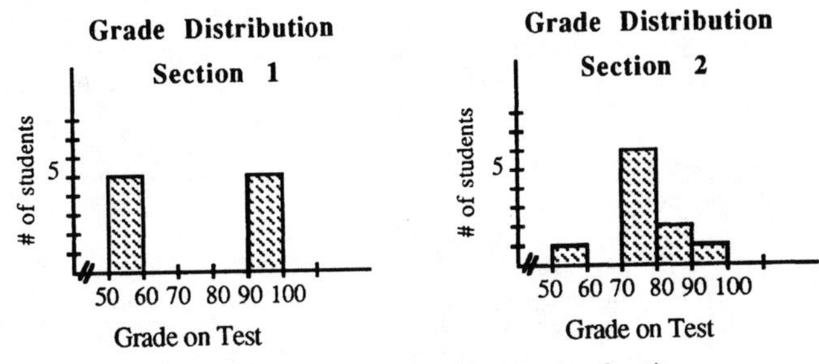

Figure 1.2 Grade Distribution by Section

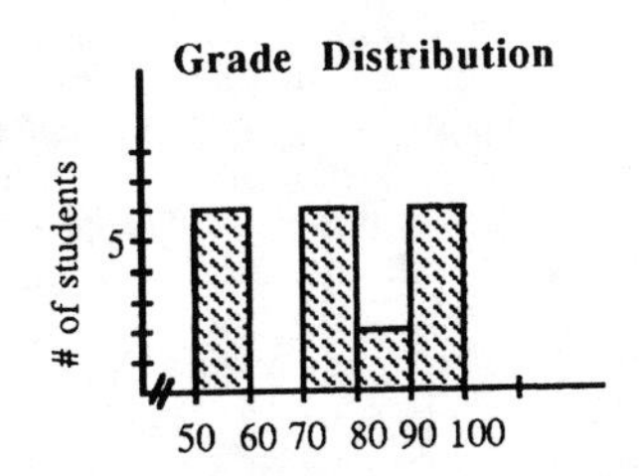

Figure 1.3 Grade Distribution (Two Sections Combined)

<u>Evaluation of Responses</u>: Before we can evaluate the responses, we must understand the question. The student asked, "How were the tests?" But, is this really what the student wanted to know? Would it be fair to assume that the student really wanted to be able to evaluate "how I did relative to the rest of the class, or relative to some standard (such as grade)?" The student has a personal interest in one particular grade, and our education system has conditioned us to focus on grades. The revised statement is probably a more accurate statement of the intended question. This points to a basic rule for data collection and reporting results:

- The purpose of the study must serve as a guide to data collection and reporting.

With the appropriate question in mind, consider the verbal response–stating the spread and average grade. The statement is mathematically correct for each class or for the combination of the two classes. (The average was calculated by totalling the grades and dividing by the number of grades.) Clearly, the statement fails to give any information about "how the grades cluster." For Class 1, the response seems somewhat misleading–no individual student made a grade close to 77. There are clearly two groups of students in this class. One group was on the same "page" as the teacher; the other group was not. For Class 2, the response is more appropriate. Some people may question including the one student who made the 54–since this student appears to be an exception. These cases point to another important point in reporting summary information:

- The availability of a mathematical formula and numbers does not make the result of the calculation meaningful.

Next, consider the separate graphs for each class. These probably provide the best description of the grades (relative to a traditional 10 point grading scale). Students will be able to look at their individual grade and evaluate "how they did."

Finally, consider the graph that summarizes the two classes. This graph may allow students to evaluate how they did relative to the larger group, but it gives little information about the specific section they are attending. From the teacher's point of view, this summarization hides information that could be valuable in planning future lectures. This graph hides the fact that there are two very distinct groups in section one. Hopefully, the teacher would recognize this, but there are many reasons why this may not happen. This display points to another important point in data collection and reporting:

- Summarizing results from two (or more) distinct groups may hide information.

Inferential Statistics

In inferential statistics a small group of items, from some larger group, is studied. The information gained from the small group is used to describe or make an estimate of some characteristic of the larger group. For example, you may want to estimate the proportion of adults living in the city who own a coffee maker. Instead of contacting every adult in the city (assuming that this were possible), you contact part of the adults in the city and determine the proportion of these people owning a coffee maker. Most business statistics books are dedicated to the study of inferential statistics. Students are introduced to methods of selecting items for a sample, formulas for determining the appropriate sample size, and the concept of confidence in the results. Your textbook will provide further study about these topics.

In this section, we will introduce some of the language of inferential statistics and illustrate these terms with examples. The terms (and concepts) to be covered include: population or universe, frame, sample, census or equal complete coverage, and sampling error. Figure 1.4 gives an overview of the relationships among some of these concepts.

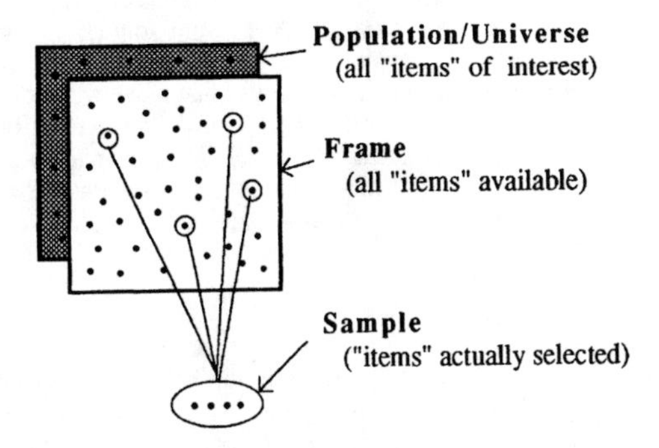

Figure 1.4 Graphical View of Inferential Statistics

As the concepts are introduced, the following example will be used to clarify these concepts.

Example 2: Suppose that you are the head of a national credit card company. You would like to know the average annual income of your current card carriers. You recognize that getting this information from every current card carrier may be difficult (and costly), so you decide to estimate this quantity by using information from a portion of your card carriers. Many questions must be answered to design a survey that will capture useful information. Some of these include:

- What do you mean by "annual income" and "current card carrier?"
- How will you identify your "current card carriers?"
- How will you decide which "current card carriers" to ask for information? (This includes two questions: how many to contact, and which specific individuals)
- How "good" is your estimate?

Population/Universe: In this material, the terms population and universe will be used interchangeably. Some statistics textbooks will use the term population; others will use the term universe.

The universe is the "conceptual" collection of all objects of interest. The word conceptual is used to emphasize that some objects may not be accessible. The universe represents all of the items that you would like to be able to study.

For our annual income example, the universe could be all individuals who currently carry a valid (not expired or revoked) card from your company. You must operationally define what is meant by the word "carry." An operational definition would describe "card carrier" in such a way that two people looking at the same individual would agree whether that individual should be classified as a card carrier or not. An operational definition should eliminate any confusion in cases such as those presented in the following cases. If a student has a credit card in their possession but the credit card is in a parent's name, who is included in the universe? Likewise, if two cards are issued in the same name, does this represent one or two card carriers?

Frame: Once you have identified the universe or population of interest, you must physically obtain a list of all objects in the population/universe (or devise some other way of actually accessing this group). This list (or other representation of the members of the population/universe) is called the frame. Therefore, the frame represents all of the items that you could choose to observe for your study. Most attempts to get a complete list of members of the population/universe will not form a perfect match. Most of the time the frame will omit some items included in the population/universe and include some items that are not part of the population/universe. Subject matter experts, not necessarily statisticians, must be used to help identify a frame that closely matches the population of interest. The "goodness" of the results of the study will be affected by the choice of frame since any statistical inference will apply to the frame. Any inference from the frame to the population/universe will be based on the subject matter expert's belief about how well the frame matches the population/universe.

In the credit card example, the billing department may have a list of "all" accounts. You might choose to use this list for the frame. Initially, you may believe that this frame is an exact match for the population/universe. A closer study would probably uncover differences. For example, individuals who have been issued cards recently may not have been added to the list (i.e., there are individuals in the population/universe that do not show up in the frame). On the other side of the coin, some people may have made a decision to quit carrying your card. Their notice to you may be "in the mail" or they may have cut the card in half and not notified you. In either case, you would still show their name in your records as a current card carrier (i.e., there are individuals in the frame that are not part of the population/universe of interest) How often is the list updated? Even if the list is continuously updated, the survey that you conduct will not be done at the instant that you receive the list–changes could occur during the interim.

Sample: A sample is a collection of some of the items in the frame. Samples can vary in size from one to every item in the frame. The sample represents all of the items that you do choose to observe in your study. The method of obtaining the sample and the size of the sample will affect your degree of confidence in any estimate about the frame. Some form of random sampling is used for most studies involving inference. The use of random samples allows us to determine how much the estimate could vary from the value that we would get if we looked at the entire frame. (You have seen plenty of reports that say that the actual value would be within 3% of the value reported in the study, or that the actual value would be within $1000 of the reported value.) You will use your textbook to study the mechanics of random sampling in some detail. Your book will also present formulas for determining the appropriate sample size. In general (when we are using random sampling), our degree of belief in the results obtained in the sample increases as the sample size increases. Of course, the cost to obtain the information increases, and our ability to maintain the same level of detail decreases.

A sample for the credit card example may consist of 100 randomly selected individuals on the list provided by the billing department. We should note that all of the statistical theory that estimates sample size and the degree of belief in the results from the sample assumes that you will be able to reach, and get a response from, each item (in this case individual) selected for the sample.

Census or **Equal Complete Coverage:** If you conducted a census, you would observe every item in the frame. A census is another name for a 100% sample. Similarly, an equal complete coverage is the result that you would have obtained if you had observed every item in the frame–using the same definitions, procedures, care, workers, and training at approximately the same time that you took the sample.

In our credit card example, a census or equal complete coverage would have contacted every individual on the billing department's list, used the same method of contacting them, made the same effort to insure an accurate response, and done this at the time that the sample was collected.

Sampling Error: Sampling error is the difference in the estimate that you receive from the sample and the result you would get from a census. Looking at a sample gives an estimate of a characteristic (such as, the average income of all card carriers). Most people recognize that another sample from the same frame might result in a different estimate for the same characteristic. Hopefully, both estimates will be "close" to the value that you would receive if you looked at every item in the frame (every individual on the billing department's list). Sampling error is a measure of how "close" the estimate is to the value you would receive from the census.

Obviously, when you are using sampling to estimate some characteristic, you will not be able to determine the sampling error for your specific sample.

(If you knew the value that you would receive from a census, why would you be sampling to estimate that quantity!) Statistical theory, developed in your textbook, allows us to estimate the expected magnitude of sampling error for certain methods of sampling.

Sampling error is one of the most misunderstood concepts in inferential statistics. Many of the "problems" that people refer to as sampling error are actually design errors. Certain problems would arise whether we took a census or a sample. For example, people will refuse to respond, people will lie, people will misunderstand the question, the answer will be incorrectly recorded, etc. These problems need to be minimized through careful design of the study. No approach to sampling can "fix" these problems, and an estimate of the effect of these problems is **not** included in the reported sampling error. A few examples will clarify some things that are not included in sampling error.

- Sampling error does not account for failures of the questionnaire (poor or missing definitions, ambiguous questions, etc).

  Recently, a professor received a survey in the mail. The survey contained $1.00 to encourage the professor to give it some attention. This worked–at least until the first few questions were read! One of these questions asked, "How many computers are in your company?" Should the professor respond with the number of computers in the department (probably known to the professor), the number of computers in the School of Business (possibly known to the professor), the number of computers in the university (unknown to the professor), or the number of computers in state agencies (unknown to the professor)?

- Sampling error does not account for the poor or inappropriate choice of frame.

  The day of the 1948 presidential election the *Chicago Tribune* went to press with a front page headline of "Dewey Defeats Truman." History reflects that this did not occur. The surveys that predicted Dewey as the winner used a list of people with telephones as their frame. In 1948, telephones were not in as many homes as they are today. Only the "well to do" had telephones. This group also tended to include more Republicans. Many people who did not own a phone did vote (these people were part of the population of interest but not part of the frame).

- Sampling error does not account for how non-response affects the results.

  Not being included in the frame and not responding to a survey are two different sources of variability in a study, but neither is included in sampling error. People who do not respond (voluntarily or involuntarily) do so for a variety of reasons. These people may have the same opinions as the people who do respond, or they may form a block of people with a single opinion. Without their response, only subject matter knowledge can help determine how these people would respond.

A recent survey of Girl Scouts attempted to determine how girls in Central Virginia viewed camping. A sampling plan was developed to contact urban and rural girls and girls from different age groups. When the survey was conducted, a large proportion of the girls living in low income urban areas failed to respond. It was knowledge about camping and the people in the areas that did not respond (not statistical theory) that allowed the survey group to conclude that the results from other geographic areas should not be applied to these geographic areas.

- Sampling error does not account for a poor choice of variables.

A few years ago University Food Service wanted to evaluate how the university community viewed their "quality of service." What constitutes "quality service?" Is it appearance, variety of entrees, taste, price, friendliness, cleanliness, speed, or some combination of these. Which of these are most important to customers and potential customers? If the survey concentrates on the wrong variables and changes the services provided based on the results of the survey, business may decrease.

- Sampling error does not account for errors in measuring, recording, and processing.

Reports of sampling error assume that the measurements reported are correct. In addition, the numbers reported, the numbers recorded, and numbers used in computations are assumed to be the same.

- Sampling error does not account for the unwillingness of some respondents to provide correct information.

Some questions, especially those that deal with people's values, touch sensitive areas of people's lives. Some people will fail to respond to these questions, some will make up an answer, some will give the answer that they believe you want to hear, and others will answer the question that was asked.

When you ask a question about personal income or expenses, the answers will often depend on the perceived use of the data. How many people report the same income to the IRS and on a country club membership application? How many people report the value of their house the same to the tax assessor and the realtor trying to sell the house?

In the recent Virginia gubernatorial election, exit polls reported a landslide victory for Douglas Wilder over Marshall Coleman. The actual margin of victory was less than 1%. Was the error in the results of the poll due to drawing a sample that was not "representative" of the voting public, or did the issues of race and abortion cause people to respond to the poll in a different way from the way they voted in the booth? There is no way to know the answer, but most statisticians (and psychologists) would bet that the sampling technique was not the problem.

- Sampling error does not account for differences that arise from using a frame from one population or universe to understand some other population or universe.

  This is the most often used erroneous interpretation of sampling error. A survey is taken or a study is conducted at some point in time to determine what will happen some time in the future. It is not reasonable to assume that the conditions that were present at the time of the study will be the same as the conditions that will exist in the future.

  Most people try to use political polls taken prior to an election to predict the outcome of the election. Attempts are made to contact people who will vote, and these people are asked, "If the election were held today, ..." Of course, the election will not be held that day, the person may honestly plan to vote and then not vote, or they may respond with their current choice for the office to be elected but change their mind before the election. The population or universe of interest are the people that will vote in the election (including the opinions they will express in the voting booth). There is no frame that will match this population prior to the election.

  Planning for future production falls into the same category. Products are produced day after day. Customers have identified characteristics of the product that are important. The producer would like to know how the product that will be produced tomorrow will stack up next to the customers' requirements. There is no frame that will allow you to sample tomorrow's production today!

Analytic Studies

Analytic studies attempt to address this last problem: how do you plan for (predict) future occurrences when you cannot sample future outcomes. Analytic studies recognize that the conditions that produced one set of outcomes will never be seen again, but that the conditions may combine in such a way that output from the process becomes predictable–within some limits.

Rather than looking at one collection of data like a "snapshot," analytic studies try to create a "motion picture" of the process over time. Subgroups, collections of several observations from a short time period, are studied to understand the past behavior of the process. If the process output is stable, and the people with knowledge of past conditions affecting this process agree that future conditions will be similar, the future output will be predicted. Figure 1.5 show the relationships described.

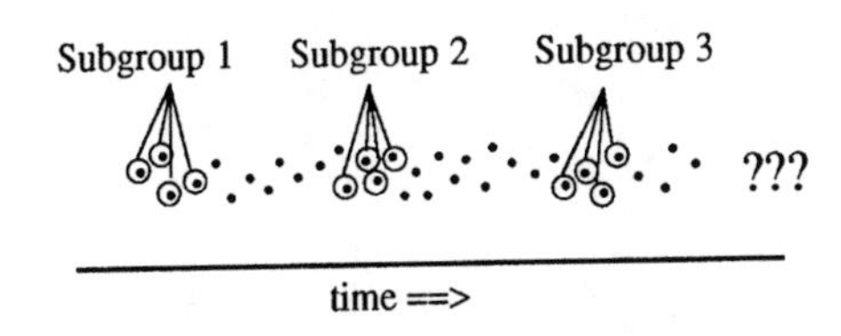

Figure 1.5 Graphical View of Process Studies

The goal of analytic studies is to learn about the process that produced the outputs in order to improve the future output from that process. To improve processes people must be able to identify cause and effect relationships, they must be able to evaluate changes to the process, and they must be able to determine the types of actions that are appropriate. Sometimes people need to question why a specific outcome is "different." At other times, people need to recognize that variation is part of the process, and reducing this variation will require fundamental changes to the process. Making this distinction requires the ability to distinguish common cause and special cause variability. The other modules will expand on the concepts of process studies and will present some tools and techniques that can be used in process improvement.

Exercises

1.1 Find an example of descriptive statistics in a newspaper or magazine. Write a short paragraph explaining what has been presented in the article or figure.

1.2 Find an example of the use of inferential statistics in a newspaper or magazine. Answer the following questions.
   a. What are they estimating?
   b. What sampling error have they reported?
   c. What is the population of interest?
   d. What frame has been used?
   e. Are the population and frame clearly explained in the article?

1.3 List three real world situations where prediction of future output from a process may be helpful (i.e., where analytic studies would apply).

Statistical thinking will one day be as necessary for efficient citizenship as the ability to read and write.

H. G. Wells

Module 2: Statistical Thinking

Many people have a fear of statistics and the study of the subject. Some of these people claim that they see no use for the material in their job or their life. Yet, these same people–without even realizing it–use the concepts of statistical thinking every day. Some simple (basic) examples include deciding what time to leave for work, deciding if they need to carry an umbrella, deciding if they should buy an item at the store, and almost any other decision that involves making a choice or planning.

Consider what these examples have in common–all are decisions that involve the recognition of variability and how variation will impact their actions. These decisions will not be made in a vacuum. Events preceding these decisions will influence the current decision, and the results of (or repercussions of) the decision will be evaluated in light of history. When people decide what time to leave for work they recognize that travel time varies from day to day, but most people (after travelling a route for a number of days) feel comfortable estimating the travel time–give or take a few minutes. To decide whether to carry an umbrella, a person may listen to the forecast and make a decision based on their experience with the forecaster, or the person may look out the window and make their own "guess" based on what has happened on days that looked similar. When deciding whether to buy a certain product the person must consider their bank account, the product itself, and their need for the product. No one doubts that the balance in a checking account varies from time to time, and the dollar amount of outstanding bills also varies. In addition, the "quality" of an item will vary from one item to another from the same vendor, from one item to another when they come from different vendors, and from one person's perception to another person's perception for the same item. After purchasing an item, the person's evaluation of the goodness of the item will be influenced by their perceived need for the item, how well the item performs, their expectations for the quality of the item, as well as their ability to afford the item.

Statistical thinking does provide a strong base for the formal study of statistics. But, this should not be the sole purpose for studying the concepts of statistical thinking. The examples included in this module will introduce you to some concepts that can be applied without the use of traditional statistical tools and techniques.

Statistical Thinking In Process Studies

Statistical thinking is the use of a collection of concepts that depend on understanding the effects of variation on the outputs from a process. People who think statistically recognize that the study of processes can influence the way we view everything, that processes produce results over time, that these results vary, and that the variability in these results may be caused by special occurrences that affect one (or a few) outcome(s), or the variability may be inherent in the process that produced the results. Distinguishing these two causes of variability helps determine appropriate action for reducing variability in future output. In addition, statistical thinking leads people to recognize that changing the results of a process involves understanding the inputs to the process and how changing these inputs (individually or in some combination) affects the results. Statistical thinking helps people understand that optimizing the outcomes does not result from optimizing each area (department, function, etc.) that is involved in producing the outcome.

Using statistical thinking effectively requires understanding the concept of a process, identifying sources of variability, acknowledging the role that measurement plays in our view of processes, insisting on the need for and use of operational definitions, recognizing the need for distinguishing between common causes (inherent to the process) and special causes of variability, and communicating the limitations of any study. These areas will be explored in more detail in the following sections and modules.

Process

A process may be something as simple as making a telephone call or as complex as building a computer. Many processes can be broken down into sub-processes; or we may choose to call a collection of interconnected processes a system. Various definitions of process exist. Webster's Dictionary defines a process as "a method of doing something, with all the steps involved." A popular definition used in the quality improvement arena is "a blending or a transformation of inputs such as people, materials, equipment, methods, and environment into outcomes.[1]" Hospital Corporation of America uses the following definition, "A process is a series of actions which repeatedly come together to transform Inputs provided by a Supplier into Outputs received by a Customer. The Outcome is the degree to which Outputs meet the needs and expectations of the Customer.[2]" All of these definitions either state or imply that there is some outcome of interest to someone. Identification of the outcomes (and outputs) of interest requires thought and communication with the customer of the process. Note that customers may be internal to the organization or external.

Any characteristic that we observe over and over again could be viewed as the output of a process. Included are such things as your monthly electric bill, your car's gas mileage, the time that a particular parking lot fills, the diameter of coils of wire produced by a plant, the number of long distance calls placed by a company each day, any number that is reported as a measurement of some dimension of a product, the number of times a computer job must be edited before it runs successfully, the number of hours spent studying, . . . Recognizing that data come to us repeatedly allows us to build on the history of the process. Planning for future outputs and outcomes becomes more rational.

Variation

One of the "givens" in process studies is that the output from a process will vary. If you do not see variability in the output from a process, consider two possibilities. One possibility is that you may not have refined your measurement device to pick up differences. For example, suppose you were given an ordinary ruler and asked to measure the length of three pieces of typing paper. Depending on the markings on the ruler, most people would report the length of each piece as eleven inches. If you were given a device that reads to the nearest 0.0001 inch, you would probably report a different answer for each page. Another reason that variability may not show up is that people are not reporting observed results (out of fear, lack of time to measure, lack of understanding the need, etc.).

Example 1: In a recent television advertisement for a grocery store four shoppers were picked "at random" and asked how much money they saved each week by shopping at that store. The first shopper answered, "I save $10.00 a week." The second shopper answered, "I save $10.00 a week." The third and fourth shoppers answered exactly the same. One of the shoppers was carrying a small basket with a few items. The other three had shopping carts filled to various levels. There are at least two reasons to question the truthfulness of this ad. First, there is no variability in the responses provided by these "randomly selected" shoppers. Four people would not be expected to save the same amount of money (to the penny). A less obvious reason to doubt the ad is based on the shoppers' ability to answer the question. How do they know how much they save each week? Do all other chains have the same prices as each other (otherwise the amount of savings would depend on which chains are being compared)?

Variability is such an expected part of life that we often fail to recognize some of the consequences. Consider the task of a facilities planner designing a university, airport, hospital or other public use facility. At a university, the number of students and faculty will vary from morning to night and day to day. Course schedules will vary (not just the time of the classes but the courses themselves). The facilities must be built to accommodate these differences. In some cases, the facilities will actually force some of these differences–there may not be enough classrooms to allow all students who want to take a class at 9:00

a.m. to be accommodated. Airports recognize that not all people on a plane will have the same final destination. As a result, the airlines develop systems to help move people, luggage, and equipment in something that resembles an orderly fashion. (One time a first time flier asked, "Why can't they just put THE connecting flight at the next gate?" The passenger soon realized that people on the plane were connecting to a variety of different flights.) Hospitals do not build facilities to serve every person in the community at the same time. They recognize that the number of emergency room visits will vary from day to day (with more visits on holidays), that the proportion of people admitted to the hospital will vary, and that the types of services required will vary.

In manufacturing, variability is so expected that most designs state a nominal (target or desired) dimension and a range of acceptable dimensions for each critical characteristic of a product. Unfortunately, many times this range is used as an excuse for not improving processes. If there are zero defects (i.e., if parts measure "in specs") the reaction has been, "If it ain't broke, don't fix it." A little statistical thinking will help explain why zero defects is a limiting goal. Figure 2.1 shows a classical view of the meaning of specifications. According to this graph, there is a target value and an interval of acceptable values for some measurement of a part. It implies that any two parts with measurements in this interval will function equally well. Parts with measurements outside this interval will result in a standard cost to the manufacturer. That is, any part inside the specifications is perfectly acceptable, and any part outside the specifications is totally unacceptable.

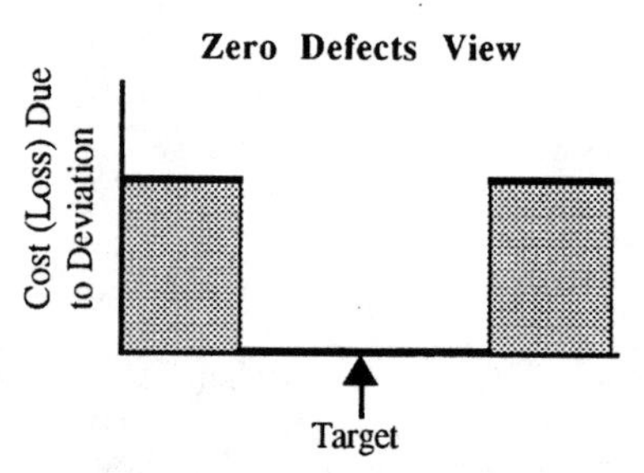

Figure 2.1 Perceived Losses from Zero Defects

Figure 2.2 shows a more realistic view of the cost of deviations from the nominal (target) value. This graph implies that parts made to nominal will cause few problems and will incur little unexpected expense. The further the part is from nominal, the more problems you expect to encounter. As problems increase so do costs.

What has just been described is the <u>Taguchi Loss Function</u>. The loss function shown in this graph is symmetric, but it is possible to have loss

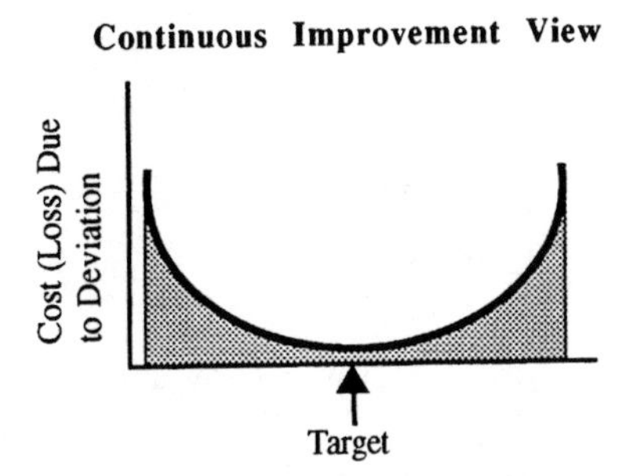

Figure 2.2 Losses from Being "Off Target"

functions that are not symmetric. For example, you may be drilling holes in a board. The hole is suppose to have a certain diameter so that a metal rod may be placed through the hole at final assembly. If the hole is slightly too small, the rod will not fit, the product will have to be disassembled, the hole redrilled, and the product reassembled (very costly). If the hole is too large, the appearance will change slightly. The rod will still fit, and the product will function (almost) as designed. The loss function will increase slowly (at first) as the diameter increases. Then at some point the curve would take a steep upward turn–representing the need to scrap the product.

Example 2: When you open a soft drink you expect the cap to be reasonably easy to remove and the drink inside to taste fresh. Many factors influence whether your expectations will be met. One of these factors is how well the cap fits the bottle. The soft drink company purchases bottles from one or more vendors. They purchase caps (or closures) from different vendors. If everything works right, the bottles are placed on the bottling line, they are filled with liquid, and the caps are placed on the bottles. This is done at a rate of several hundred bottles per minute. The bottle makers and the cap makers have specifications on a number of different characteristics. One of these characteristics is called "E dimension." For bottle makers, the E dimension is from the outside edge of the bottle on one side to the outside edge of the bottle on the other side. E dimension on the cap is measured inside the cap from the point of the thread on one side to the point of the thread on the other side. The bottle and cap are shown in Figure 2.3.

For makers of 16 oz. glass bottles, the target E dimension is 0.984 inches. The specifications allow for E dimensions between 0.972 inches and 0.997 inches. For makers of the cap, E dimension has a target of 1.003. The specifications allow for E dimensions on the caps to range from 0.995 to 1.011. If the bottler plans (and predicts their operating costs) using a zero defects approach, they will expect to see the line running smoothly as long as

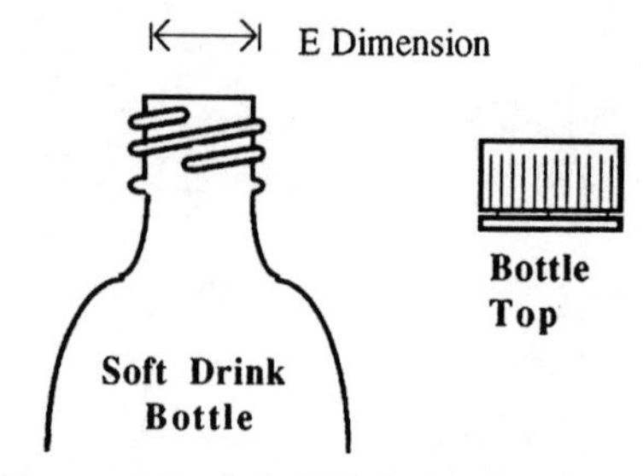

Figure 2.3 Soft Drink Bottle and Top

all bottles are in specifications and all caps are in specifications. If we write both sets of specifications together (as seen in the table below), we will see that zero defects will not guarantee smooth operation.

	Target	Lower Spec	Upper Spec
Bottle	0.984	0.972	0.997
Cap	1.003	0.995	1.011

Since the diameter of the cap MUST be larger than the diameter of the bottle for the cap to fit onto the bottle, we can expect to have problems when we try to use a large bottle and a small cap (both within specs). On the other hand, if we have a small bottle and a large cap, the drink inside may not stay fresh. Producing more parts (both bottles and caps) closer to the target dimension will reduce the losses caused by the variation in the parts.

Recognizing variability when you see it is only a start to using the concepts of statistical thinking. In addition, you must be able to identify sources of variability. What caused, or could cause, variation in some observable phenomenon? Possible sources of variability include methods, materials, machines, people, the environment, and measurement. Each of these sources can be expanded. For example, suppose that we are trying to understand why a company phone bill is "high." Some methods that may need to be considered include job training (do employees know when they should be using the phone and how to use special features such as WATS), the personnel evaluation system (are people rewarded for making a lot of contacts), and the procedure for selecting suppliers (for phone service and for other supplies). Changes in any one of these methods may result in changes in the phone use patterns and in many other places. Module 5: Cause and Effect Diagrams will introduce a technique that can be used to help identify some of the sources of variability in the outputs from a process.

Common Cause/Special Cause

Variability in a process can be divided into two categories: common cause and special cause. Common cause variability results from the multitude of sources that affect every outcome from the process. Common cause variability is inherent in the process. A fundamental change in the process is required to reduce common cause variability. Special causes are assignable causes of variation that affect one or a few outcomes from a process. When a special cause is detected, immediate action should be taken to attempt to identify the source. Special causes of variability that reoccur in a predictable manner may be a signal that what was thought to be a single process may be a combination to two or more processes.

Example 3: This example was used by Walter Shewhart, an early pioneer in the field of quality control, to demonstrate the difference in the two types of variability. He suggested the following:

- Take a pencil and piece of paper.
- Write a lower case letter "a."
- Study that "a."
- Make another "a" just like the first one.
- Make a third "a" just like the first two.
- Look at your three "a's."
- Circle the one you like best.
- Think about what you did differently when you made that "a."
- Make another "a" just like your favorite.
- Put your pencil in the other hand.
- Make an "a" just like the other four.

Most people agree that the first four "a's" look more alike than the last one. But, they do not look exactly alike. There are a variety of reasons for the differences, but each reason affects all four results to some degree. These results came from the same process and should not be ranked (for the purpose of rewarding the one on top or punishing the one on bottom). Only common cause variation was present in this process. Improving the "a's" produced by this process would require a change in the process–such as using smoother paper, or a different shaped pencil. The last "a," the one made with the other hand, is clearly different from the others. Something special happened to the process when this "a" was produced. It should be singled out and questioned.

People who understand the distinction between common cause variation and special cause variation are at an advantage in the business world. It is this distinction that helps direct improvement efforts in a process, and the failure to understand this distinction can cause problems.

Example 4: Campbell's Soup has observed the sales of one of their products over a long period of time. Figure 2.4 shows approximately what has happened.

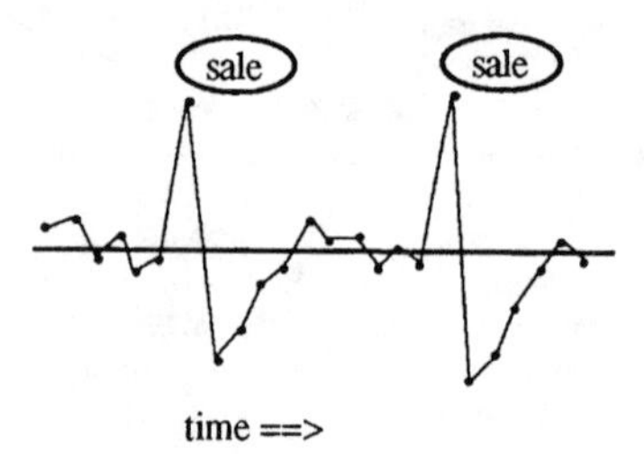

Figure 2.4 Effect of Sales (When Demand is Stable)

Sales for this product varied from week to week. Other than during sales and just after sales, the demand for this product was fairly stable. Between sales orders varied from week to week for a number of different reasons, but the combined effects of all the causes was predictable (in terms of the number of orders to expect). When Campbell's offered a sale on this product, companies (just like individuals) "stocked up." When the sale ended sales dropped to below their pre-sale levels. You may say that all of this is predictable–and it is. The interesting thing is how Campbell's used statistical thinking to analyze the situation. They reasoned that as long as the process was stable they could predict (within a range) the demand in coming weeks, order ingredients and supplies accordingly, schedule personnel and equipment, and maintain a fairly steady work flow without incurring high inventories. By offering "sales" they were upsetting their own operations, causing overtime during sales and idle time for the next few weeks, making procurement of ingredients more difficult, and increasing costs. If they quit having these promotions they could lower the regular price of this product and increase their profit! (Procter and Gamble used similar logic when they announced a cut back on the number of coupons they would print.)

Processes are studied using Run Charts (Module 8) and Control Charts (Module 9) to help determine when special causes of variation are present in a process. Even without the knowledge to set up these charts, you should recognize that reducing common cause variation is the responsibility of management (with the help of workers in the process). The manager of the process is usually the only one with the authority and resources to make fundamental changes to the process (like how supplies are purchased, how new employees are trained, when maintenance is performed, etc.). Special causes of variability are usually local in nature. The workers in the area will have the most information about what was different at the time this output occurred.

Information about output from the process must be provided close (in time) to when the product was produced if you expect to be able to track down any special causes. The workers may need the assistance of management to remove the special cause.

People who react to each individual observation from a process as though it is a special cause, and make adjustments based on this observation, may be tampering (and increasing the variation in the process). If you doubt this, try the "Plinko" game suggested at the end of this module.

Measurement

One source of variability that is often overlooked is in the area of measurement. You hear the announcement telling the number of people attending a college football game or you hear reports about the number of cars that Ford (or some other car maker) sold last month, and you think you <u>know</u> something with certainty. If someone else counted the same thing, would they get the same answer; if the same person counted again, would they get the same answer. The answer is probably not.

Operational Definition: For results from any measurement process to have meaning, terms must be operationally defined. An operational definition describes what is to be measured and the steps to be followed to measure it. If the results of the measurement are to be used to evaluate the characteristic of interest, a decision criteria must be established. Terms such as round, square, soft, on time, sharp, and clean have different meanings to different people. For customer's and supplier's to do business with these terms they must come to agreement on operational definitions.

Example 5: Good operational definitions do not guarantee perfect results as this example from the American Society for Quality Control[3] will show. They present the example to show the fallacy of using 100% inspection to insure quality of product. The results should make you think about any measurement. You may want to try this experiment with several people to see how much variability you can see from such seemingly straight-forward instructions.

To give you an idea of just how difficult visual inspection can be, take this quick test. Count the number of times the letter *f* appears in the next paragraph. Give yourself two minutes to finish (inspectors work under time constraints), and don't mark the page (inspectors cannot mark on the product). You will find the answer at the end of this module (before the exercises).

"The necessity of training farm hands for first class farms in the fatherly handling of farm live stock is foremost in the minds of effective farm owners. Since the forefathers of the farm owners trained the farm hands for first class farms in the fatherly handling of farm livestock, the farm owners feel they should carry on with the former family tradition of training farmhands of first class farms in the effective fatherly handling of farm live stock, however futile, because of their belief that it forms the basis of effective farm management efforts."

Now, what happens if you add some kind of measurement device such as a scale, watch, or ruler. For example, you weigh something, time something, measure the length of something, etc. Will the measurement device return the same value when used by one person to measure the same item repeatedly? What happens when the same measurement device is used by multiple people to measure the same item repeatedly? If you have multiple measurement devices that are used by multiple people to measure the same item over and over again, will you get the same answer each time? If you do not get the same answer (and this situation is not uncommon), which answer do you believe?

Most people forget that any number that is reported is the result of a measurement process. If you change the method of measuring, you may have a new number to report. The "true value" of any characteristic of interest will never be known since any reported number will be dependent on the measurement process and the definitions used by the individual doing the measuring.

A measurement device that provides consistent results over time is a prerequisite to comparing different pieces of the "same" product. If you cannot trust your measurement device, how will you be able to tell if two different measurements are the result of differences in the products or your inability to get the same answer twice!

In addition to providing consistent results, your measurement device needs to be able to detect differences that are meaningful for the situation that you are studying. The size difference that you want your measurement device to be able to detect will depend on the application. You may be content to know the distance to Washington, DC to within a mile, but a golfer would not be satisfied to know the distance from the tee to the hole to within a mile!

Limitations

Statistical thinking helps people recognize that all studies have limitations. There are sources of variability that are not identified, there are causes of variation that are almost always seen together, or there are things that go wrong in a study. Statistical studies must be viewed with a "dose" of subject matter knowledge. Statistics will not prove a cause and effect relationship, they will only provide support for the existence of such a relationship. Three examples will help explain how statistics can be misleading without subject matter knowledge.

Example 6: Someone decided to collect data on the number of traffic accidents each day from mid-May through mid-July and the amount of ice cream sold over the same period of time. The data for both activities increased slowly at first, spiked up during the last weekend in May, levelled off again, and spiked up during the first week in July. The figures definitely appeared to move together. Without subject matter knowledge, you might conclude that eating ice cream causes traffic accidents, or that traffic accidents causes people to eat ice cream. How ridiculous! A little thought will reveal that both spikes in the data occurred on holidays (Memorial Day and July 4). The increase at the beginning of the time could be due to a third variable–the approach of summer when more people travel and more people eat ice cream.

Example 7: One grocery chain (referred to as "Chain A" in this material) makes the following claim in their television advertisements. "Here's more proof you pay less for groceries because of 'Chain A' no matter where you shop. At 'Chain B' stores in Atlanta, where there are no 'Chain A' stores, 'Chain B' shoppers pay $198.87 for 111 grocery items. But, in Charlottesville where there are 'Chain A' stores, the exact same groceries only cost 'Chain B' shoppers $171.40 ($27.47 less). Now if 'Chain A' can save 'Chain B' shoppers that much at their store, imagine how much they'd save at 'Chain A'."

The ad implies that the presence of their stores causes the competition to lower prices. Does the general "cost of living" in a community have any effect? It also implies that 'Chain A' has even lower prices. No data is supplied to support this claim.

Example 8: Chrysler ran the following television advertisement during Spring 1991. "Ford, GM, and Honda will not have air bags in all their cars until 1994. By that time approximately 60,000 will have deployed in Chrysler built vehicles. (pause) Advantage Chrysler." While you are hearing these words, fine print notes that certain models of Chrysler cars do not have air bags.

Someone using statistical thinking to evaluate this report would want answers to some questions. A few questions would include:

- When will Chrysler have air bags in all of their cars?
- What proportion of Chrysler, Ford, GM, and Honda cars have air bags?
- What proportion of accidents (from each maker) are serious enough to need an air bag?

It may not be possible to answer the last question in the previous example. This highlights a limitation of many studies that is often overlooked. Some of the most important figures, for decision making purposes, may be "unknown and unknowable." For example, how do you quantify the effect of a happy employee or a dissatisfied customer? How do you determine the "savings" that result from an improvement that results in fewer defects at an early stage of an assembly line? You can quantify the reduction in scrap and rework at that stage, but how will the improvement change the attitudes of the workers at the next stage, how will setup times at successive stages be affected, and how will relations with suppliers and customers be affected?

Answer to the "f test:" There were 48 f's in the paragraph. To improve, you may consider turning the page upside down. The task was to count the f's– not to read the paragraph.

Exercises

2.1 Watch TV, listen to the radio, or read a newspaper. Find examples that show a recognition of variation or sources of variability. (Or where there should be a recognition of variation but is not)

2.2 Write operational definitions that could be used to perform the following tasks.

a. count the number of students at this university
b. count the number of chocolate chips in a Chips Ahoy cookie
c. record the number of hours you watch TV for a given day
d. record the amount of sleep you got in a specific day (24 hour period)
e. count the number of rooms in a house
f. weigh a car
g. measure the length of a car
h. count the number of free parking places on campus

2.3 Have five people take the "f test" in Example 5. Report the results.

2.4 "Plinko" is a game played on The Price is Right. In this game, the contestant is given three wooden disks (similar to hockey pucks). The game board resembles the one shown in Figure 2.5.

The contestant is instructed to place one disk at any of the positions indicated at the top of the board. When the contestant releases the disk, it falls through the maze of pegs eventually landing in a bin at the bottom of the board. The number below the bin indicated the number of dollars the contestant wins.

After observing what happens when the first disk is dropped, the contestant repeats the game with the second disk. The starting point may

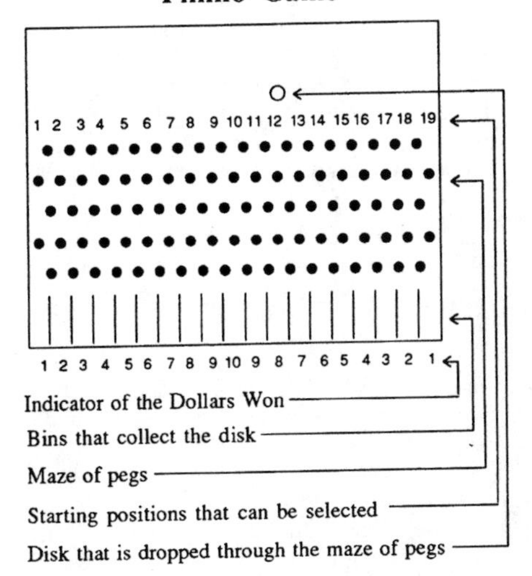

Figure 2.5 Plinko Game Board

be changed. After observing the result of dropping the second disk, the contestant repeats the game with the third disk.

Obviously, the contestant wants to maximize the total dollars won. There are several strategies the contestant could use. Three are listed below.

Strategy 1: Always drop the disk from the same position.

Strategy 2: Each time you drop the disk remember where you started. Observe the result. Make a compensating adjustment before dropping the next disk. For example, suppose that you decided to drop the disk from position 12. When you released the disk, it followed the path indicated in Figure 2.6.

The disk landed three bins to the right of the desired position. To compensate, you make your next drop from position 9 (three positions to the left of this drop). You make similar adjustments after each drop.

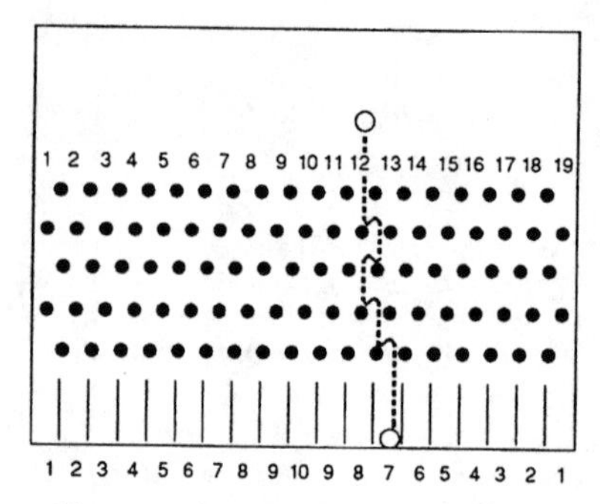

Figure 2.6 Plinko Sample Drop

Strategy 3: Each time you drop the disk, drop it from directly above where the last disk landed. (After all, "For every action, there is an equal and opposite reaction.")

a. Which strategy is best? There are two questions to address here. First, determine where the initial drop should begin. Then, determine how to choose the starting point for each successive drop. (Hint: for the second part you may want to consider what will happen after many trials–not just three.)
b. Devise a way to demonstrate that your answer is correct. Be sure to state any assumptions.

2.5 Applications of the concept presented in this game appear in many real world situations. In manufacturing, some pieces of equipment use "automatic compensating devices." After each item is produced, these devices measure some characteristic of the item and adjust the machine setting(s) before the next item is made. In education, most students study for a test, observe the result, and adjust their study habits based on the test result. List four other examples.

---

1 Scherkenbach, William W., *The Deming Route to Quality and Productivity: Road Maps and Roadblocks*, CEEP Press Books, 1988.

2 "HCA Quality: Focus on Continuous Improvement Leaders' Workshop," Hospital Corporation of America, 1989.

3 *Quality Illustrated*, American Society for Quality Control, 1985.

The long-range contribution of statistics depends not so much upon getting a lot of highly trained statisticians into industry as it does in creating a statistically minded generation of physicists, chemists, engineers, and others who will in any way have a hand in developing and directing the production processes of tomorrow.

W. A. Shewhart and W. E. Deming

Module 3: Introduction to Process Studies

Process studies can be used to understand the current operation of a process (in manufacturing, in service, or in personal lives), to plan for the future operation of a process, or to evaluate changes made in a process. The fundamental concept underlying process studies is the recognition that a combination of "elements" come together repeatedly to produce some output. These elements may include people, equipment, methods or policies and procedures, materials or supplies, and environmental conditions. Even though we may attempt to combine these elements in the same manner time after time, the results vary. Figure 3.1 represents this concept.

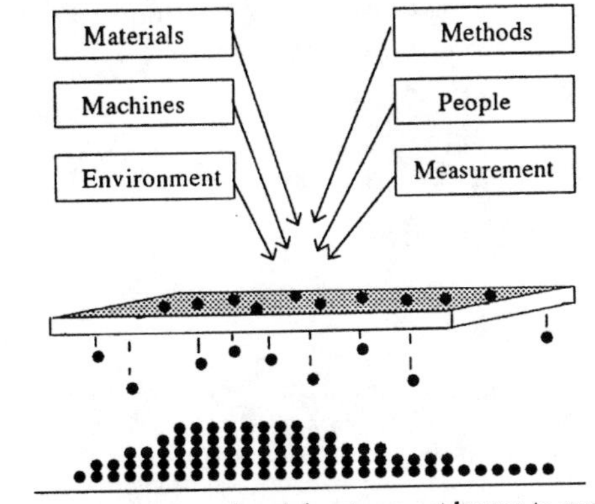

Outcomes vary even though inputs are not known to vary.

Figure 3.1 Process Overview

If we can develop ways to understand and control the variability in the output, we will be able to predict future needs and expenses more reliably. The first step to better planning requires recognizing that most numbers that we use to evaluate a company come to us on a regular basis (daily, weekly, monthly, quarterly, or annually). Each of these numbers is the result of many actions; each of these figures represents a single measure of the output from some process. We must begin to view these figures collectively (taking time into account) rather than focusing on the most recent figure or on how the most recent figure compares to one other figure (the previous one or the same figure one year ago). Once we recognize that these figures are the output of some process we can begin to identify inputs to this process and how changing these inputs will affect outputs.

Any view of a process should recognize that there are suppliers to the process and customers of that process. Understanding and improving the operation of the process requires including both customers and suppliers in any analysis. Without input and cooperation from these groups, improvement efforts may be seen as "change for the sake of change." In manufacturing, production is often viewed as an operation or a function. If we expand that view to recognize that production is a collection of processes that function together as a system (as in Figure 3.2[1]), potential improvement efforts will be expanded.

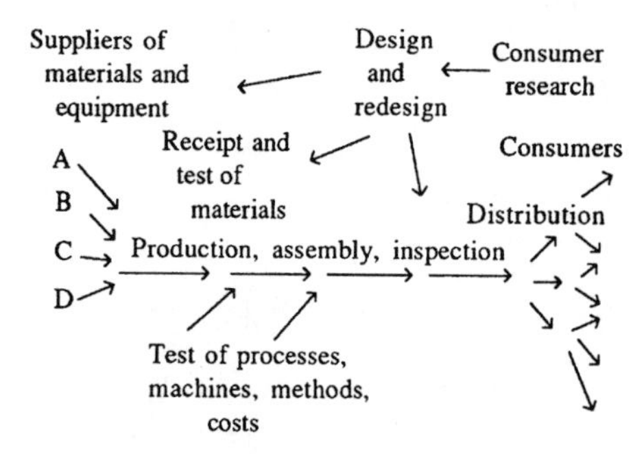

Figure 3.2 Production Viewed as a System

In this view of production we must recognize that there are many customer-supplier relationships. A customer, in the context of process improvement, is the receiver of some product or service. Customers may be internal or external to the organization. Examples of internal customer-supplier relationships may include (customer listed first) manufacturing and purchasing, shipping and manufacturing, and subordinate and supervisor. The external customer may be viewed as the person or organization who writes the check to the company for goods or services produced by the company, or the external customer may be viewed as the end user of the product or service produced by the company. For example, suppose you work for Sony in the manufacture of car radios. Sony sells some of their radios to Honda. Is Honda the external customer, or is the individual car buyer who buys the Honda with a Sony radio the external customer? Both must be considered external customers when evaluating the "radio making process."

Coordinating the expectations of internal and external customers can be difficult. Consider the education system (shown in Figure 3.3).

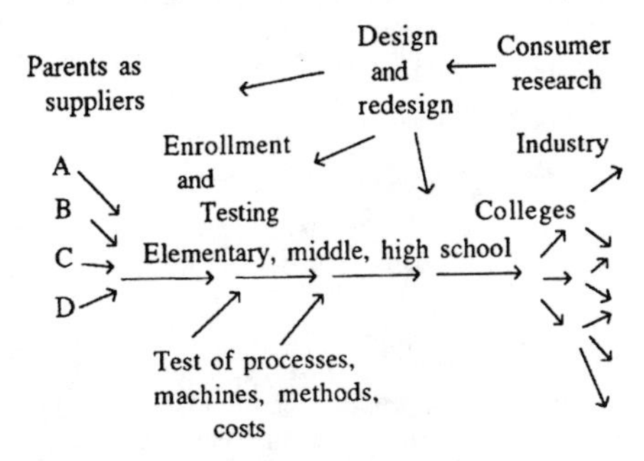

Figure 3.3 Education Viewed as a System

Some of the internal customer-supplier relationships (customer listed first) include middle schools and elementary schools, colleges and high schools, and the students and teachers. Society in general, or industry to be specific, functions as the external customer for this system. Almost daily, the news media provides feedback from the external customer (most of it very negative). According to these reports students are not learning to reason, they have not mastered the basics of geography, math, and science, and they are not learning skills necessary to function in a global marketplace. Even with these reports, students (internal customers in the system and the product for the external customer) complain about taking too much math and science. How do you reconcile the expectations of both the internal and external customer? No easy answers exist.

Voices

Analyzing a process requires the recognition that there are a variety of ways that the process can "talk" to us (just as there are a variety of customers who want to influence the operation of the process). These different forms of feedback are often referred to as "Voices." We will consider the Voice of the Customer, the Voice of the Product, and the Voice of the Process. These "voices" are depicted in Figure 3.4.

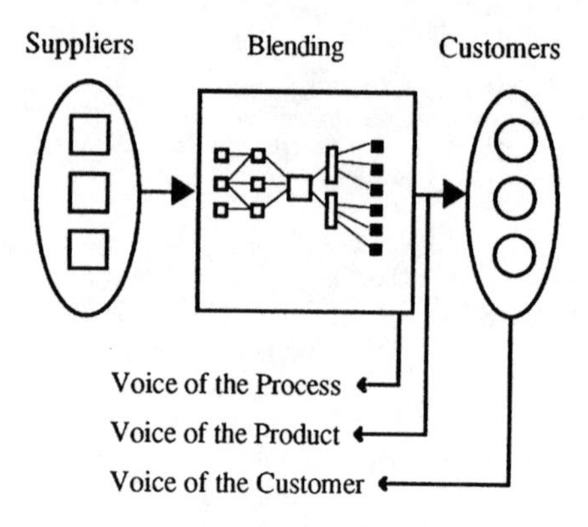

Figure 3.4 Voices of the Process

The Voice of the Customer may be heard as the customer receives and uses a product or service. The feedback may be formalized through some kind of market research, or it may be informal–such as verbal complaints to a salesperson. The Voice of the Customer is usually many different voices. Different customers find different characteristics of a product or service important; different customers have different experiences with the product or service; and different customers would have different opinions of the same product or service. In order to use the Voice of the Customer as feedback to evaluate the product or service provided by a company the information must be specific and timely. Responses like "didn't work," "missing parts," and "wrong color" do not provide sufficient information for design of future products/services or change of the current product/service.

The Voice of the Customer should be heard in the design stages for any product or service–what does the customer want or need? At the design stage two words of caution are in order. First, the customers may not know what they want. This is especially true for products and services that are on the "cutting edge." Second, what customers say and what they mean may be two different things. For example, the customers may say they want a pencil that stays sharp when they really want a writing implement that frees them from the pencil sharpener and allows them to correct mistakes.

All of the characteristics that the customer identifies as important must be translated into product characteristics. For example, the customer may say they want a desk with a drawer that holds a standard file folder. This would translate into specific inside dimensions for the drawer. As these desks are made, the inside dimensions of the drawer will vary from one drawer to the next. Collectively, these measurements form the Voice of the Product. Bringing the

Voice of the Product and the Voice of the Customer in line with each other will increase the proportion of customers who are satisfied with the product.

In order to change the Voice of the Product (in any lasting way), we must understand how the process combines inputs to form the product. This requires understanding the Voice of the Process. This voice deals with how the inputs vary and how changes in these inputs affect the operations of the process.

Example 1: Suppose you work for the maker of the plastic bottle cap found on soft drink bottles. The Voice of the Customer may indicate that people want a cap that is clean, easy to remove, and holds in the flavor of the drink. You may decide to translate "easy to remove" into a variable that measures the pounds of torque required to remove the cap. After some experimentation you find that most customers have about the same opinion about how much torque should be required to open a bottle. This Voice of the Customer can be compared to what is happening in the manufacture of the product. Experiments using caps that you produce will probably reveal that the torque required to remove the caps varies over some interval. These results will provide information about the Voice of the Product. In order to evaluate this information you will need to know how this Voice of the Product is changing over time. You will be interested in the width of this interval as well as how the values in this interval compare to what the customers have indicated that they want. At least three sources of variability will need to be addressed. Figure 3.5 will be used to help describe these three components of variability.

Effort Required to Open Bottle

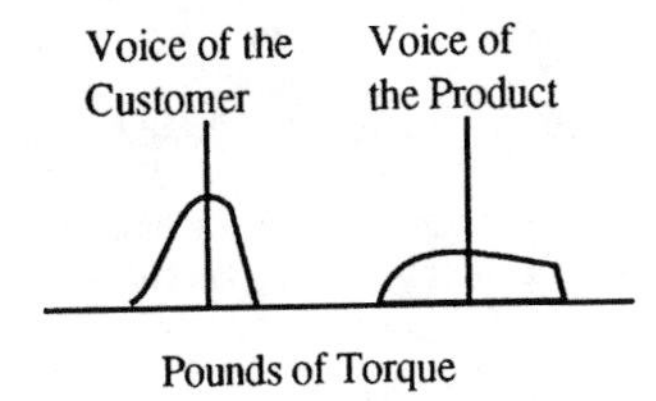

Figure 3.5 Desired vs. Actual Torque

Most obvious is the difference between the Voice of the Customer and the Voice of the Product. Customers want a cap that is easier to remove. If this were the only characteristic of interest, the solution would be easy–make caps that are a little larger or remove the tamper evidency band. Of course, as a manufacturer you must evaluate this desire in light of other characteristics of interest. A larger cap may not stay on the bottle, or it may leak. The tamper evidency band provides some measure of assurance to the customer that the product has not been altered in some unwanted way between bottling and purchase.

The two other components of variability are similar (at least in description). Customers vary from one to the next in terms of what they want, and caps vary from one to the next in terms of how they function. Reducing variation in what customers say they want may require education, or it may lead to some type of product modification. For example, many medications come in different "forms" (capsule, tablet, liquid, child proof container or not) to accommodate different customer voices. Reducing variation in how the caps function requires knowledge about why this variability exists. The Voice of the Process helps us learn about the reasons. For example, workers may study how changes in type of resin (the plastic used to make the caps), changes in the amount of resin used, or changes in the length of time the mold stays closed affect the torque required to remove the cap. Of course, variability in the size of the bottle is also a cause for concern (but this is outside the immediate control of the maker of the cap). If the key processing factors can be identified, these factors can be monitored closely to insure that the Voice of the Product will be predictable.

Why would a business student be interested in such "engineering" studies? The answer is quite simple. Most business students plan to function in some role in the management of a company. An important part of the job of management is to plan for (and insure the existence of) the future of the company. Planning for the future requires knowing the customer(s) and their needs and knowing how the company will be able to provide for these needs. Customers tend to expect products to improve over time. Improving products requires improving the processes that produce the products. Improving processes requires a team effort–most improvement efforts cut across functional boundaries. Even changes that initially appear to be contained within one department or area will probably have some effect on the work in other areas. (For example, the changes that you are seeing in your Business Statistics course should affect the way that you think about some of the topics in other subjects.) People must be free to participate on these teams, and suggestions for improvement must be supported. Management must be involved in these efforts by participating directly, providing resources, setting direction, and creating the organizational structure and atmosphere that will allow teams to prosper.

Overview of a Process Study

Even though each process will have individual characteristics, there are some basic steps for studying processes. This section will provide a summary of these steps. The following modules will present tools and techniques to support the steps discussed.

1. **Identify a process to study**. This does **not** equate to "find a problem to solve." (Problem solving can be equated to fire fighting.

Process improvement should include prevention.) Part of the process identification step involves communicating who will benefit from the study.

2. **Form a team to perform the study.** This team should include individuals who work in the process, someone who can provide the customer's perspective, a facilitator, and someone who has the resources to implement changes.

3. **Define the boundaries of the process being studied.** Everyone on the team needs to be working on the same process. Boundaries consist of the beginning and ending steps in the process. Be sure that these boundaries are operationally defined. For example, a group of hospital administrators identified the time that a patient was discharged as the beginning of the process of cleaning a room between patients. In the team of five people, there were three different interpretations of what discharged meant!

4. **Document the current operation of the process.** The existence of a standard operating procedure does **not** suffice. Have the team members develop a Process Flow Diagram (Module 4). At this point in a study, it is not unusual to discover that two people with the "same" job do this job in completely different ways. Before continuing, identify and implement a current best method for this process. (For most processes, if teams stopped at this point tremendous progress would be noted.) Another common realization is that the team assembled lacks knowledge about some part of the process. This may be an indication that the team does not include the right group of people.

5. **Identify the customers and suppliers to the process.** Determine which characteristics of the product or service are important to the customer. Determine how you could measure each characteristic.

6. **Identify inputs to the process that cause the output measure to vary**–for each output of interest. Cause and Effect Diagrams (Module 5) can help. Which one(s) of these inputs does the team believe have the largest effect? How can they be measured?

7. **Develop a plan to collect data** (Module 12). Check Sheets (Module 6) or some other method of recording the data should be devised. Before you start a large data collection process, make a trial run with the same workers. Extra time spent in designing the data collection process will pay off. Be sure to collect data over a long enough period of time to see the effect of different days, different shifts, different batches of materials, etc.

8. **Analyze the data collected.** Are the outputs and inputs stable. Use Run Charts (Module 8) and Control Charts (Module 9) to assess

stability. If special causes of variability are detected, attempt to identify these causes. If the effect is negative, work to remove the cause; if the effect is positive, work to incorporate this into the process. If the processes are stable, you can summarize the results obtained from your data with Pareto Charts (Module 7) and Histograms (Module 10). You can show the relationship between an input variable and the output characteristic with a Scatter Diagram (Module 11).

9. Identify potential areas for improvement using the information about the current operation of the process. Develop a list of possible improvement activities. Evaluate these suggestions, and select one to try.

10. Use the PDSA Cycle (Plan-Do-Study-Act) described below to implement the suggestion selected. As the team goes through the cycle, they will usually find other areas for improvement. Generally, this will lead to additional revolutions through the PDSA Cycle.

The PDSA Cycle

The PDSA Cycle (also known as the Plan-Do-Check-Act (PDCA) Cycle, the Deming Cycle, the Shewhart Cycle, and the Learning Cycle) provides a framework for scientific study. The term "cycle" emphasizes that the process continues (new learning builds on previous learning). All four stages are needed in the learning process. Figure 3.6 illustrates this cycle. The "plan" stage provides direction for the "do" and "study" stages. In the "plan" stage, questions are developed, improvement efforts are planned, and methods for collecting data to answer the questions are developed. The "do" stage will carry out the plan developed earlier. The "study" stage will evaluate the effects of the "do" stage. The actions taken in the "act" stage are dependent on the findings in the "study" stage.

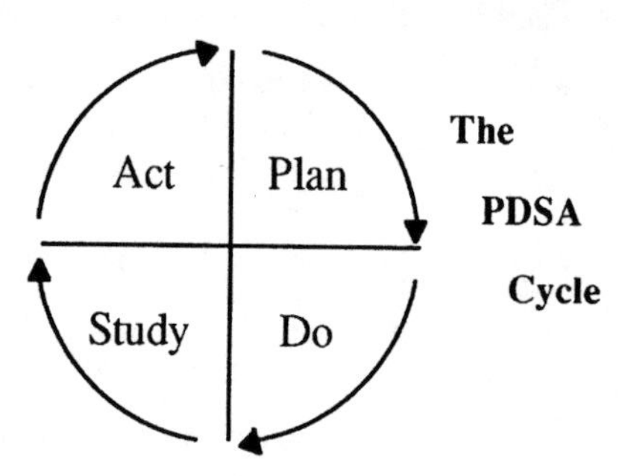

Figure 3.6 The PDSA Cycle

Each of the tools presented in the following modules will be presented in a PDSA format. The "plan" stage will include an explanation of when the tool is appropriate and instructions for using the tool. In the "do" stage, the tool will be used in a familiar situation. The "study" stage will explain how you would evaluate the results of using the tool and present any cautions that are appropriate. The "act" stage will discuss the types of actions that might follow from the use of the tool.

Exercises

3.1 Choose one of the following processes and use that process to answer the questions below.

- the registration process (for classes at the university)
- the process of getting to school each day that you have classes
- the process of buying groceries
- the process of cooking a meal
- the process of writing a research paper

a. List (in order) the steps of the process.

b. Identify inputs to the process (people, materials, methods, machines, and environment)

c. Identify three outcomes that may be important. For each outcome, determine how it could be measured.

3.2 Identify some product or service to use in the following.

a. Identify three characteristics of the product or service that may be important to a customer. Would all customers have the same expectations about the appropriate level (measurement) for each of these characteristics? If not, why not?

b. For each characteristic listed in part a, identify how the characteristic could be measured.

For example, a customer at Arby's may want to receive the right food, quickly, and be treated courteously. This might translate into the same items (made by set recipes) on the tray as on the register tape, a measure (in seconds) of the time required for service (with some determination about what is viewed as "quick"), and an employee that follows specific steps to greet the customers, take the orders, and thank them.

3.3 Consider the education process shown in Figure 3.3.

a. What role do students play in this process?

b. Who is the customer of the process?

c. How is the voice of the customer heard in this process?

d. What are some examples of inspection and rework in this process?

---

1 Deming, W. Edwards, *Out of the Crisis*, Massachusetts Institute of Technology Center for Advanced Engineering Study, 1986.

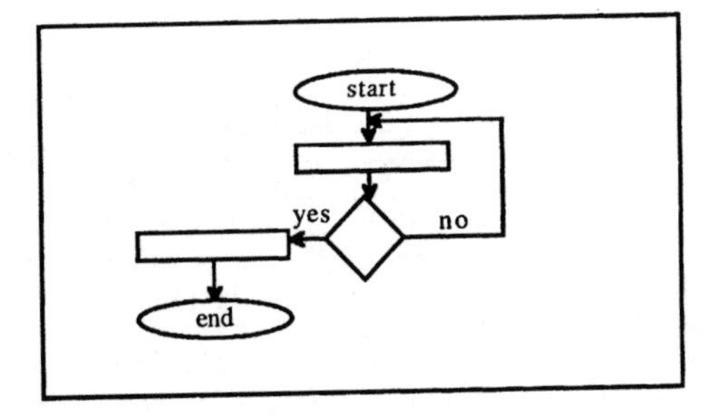

Module 4: Process Flow Diagrams

Process Flow Diagrams provide a graphical representation of how the steps of a process fit together. "Flow Charts" have been used for many years by computer programmers, but their value in process studies is often overlooked or underestimated. Process Flow Diagrams can be used in a number of ways to help reduce variability in a process. When a group of people perform "the same job," the process of developing a flow diagram will identify differences in the way the job is done and areas that lack operational definitions. Once the group (and the company) agree on a common way to view the process, the diagram can be used as part of the training program for new employees.

An early step in the analysis of any process is to understand the current operation of the process. The Process Flow Diagram offers a clear way to document the current knowledge of the process. As a team develops the diagram, they may find steps that no one on the team understands, or they may find areas where the team disagrees about how the process currently operates. In the first case, there is evidence that the proper team has not (yet) been identified. In the second case, the team needs to observe the process, determine what is currently happening, and if needed standardize the current operation of the process.

When changes to a process are suggested, modifying an existing Process Flow Diagram to show how the change would affect the process can help in the explanation and evaluation of the suggestion. This helps clarify everyone's knowledge about the suggestion. If the suggestion is adopted, the Process Flow Diagram provides documentation for training.

The diagram of "Production Viewed as a System" (Figure 3.2) is a special form of a Flow Diagram. In this case, the diagram shows how the different parts of the organization work together to accomplish the aim of the system.

Construction

Before any work can begin, the group must agree on the beginning and ending points of the process. After this is accomplished, the level of detail must be discussed. You want to include enough detail to show decision points in the process, but you do not want so much detail that you lose sight of the overall process. When you start to draw the Process Flow Diagram, you will choose the shape of the symbols to indicate the type of activity that takes place at that step. Although many different symbols are available, a few basic ones will usually convey the meaning. As you develop Process Flow Diagrams to aid in an improvement process, you must remember that clarity is important–do not add complexity for the beauty of it! The basic symbols will be described below.

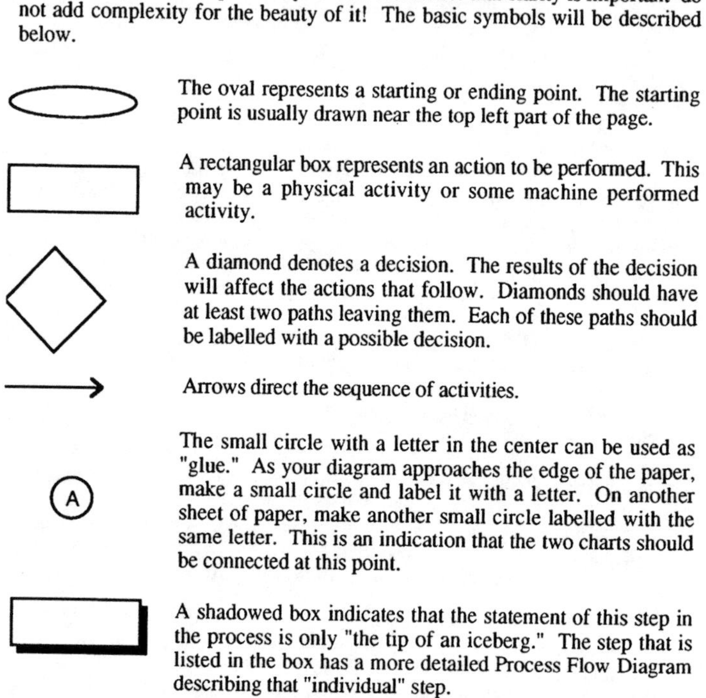

The oval represents a starting or ending point. The starting point is usually drawn near the top left part of the page.

A rectangular box represents an action to be performed. This may be a physical activity or some machine performed activity.

A diamond denotes a decision. The results of the decision will affect the actions that follow. Diamonds should have at least two paths leaving them. Each of these paths should be labelled with a possible decision.

Arrows direct the sequence of activities.

The small circle with a letter in the center can be used as "glue." As your diagram approaches the edge of the paper, make a small circle and label it with a letter. On another sheet of paper, make another small circle labelled with the same letter. This is an indication that the two charts should be connected at this point.

A shadowed box indicates that the statement of this step in the process is only "the tip of an iceberg." The step that is listed in the box has a more detailed Process Flow Diagram describing that "individual" step.

There are probably as many different approaches to drawing Process Flow Diagrams as there are individuals making these diagrams. Some people like to list the activities in outline form. Then they start to draw. Other people use "Post It" notes and a wall. Still others draw from the start to the first decision step. Then they select one of the possible paths and continue. Finally, they return to the other possible decision(s) and complete them.

Examples

There are several types of Process Flow Diagrams. The simplest form provides an overview of the process with almost all details suppressed. The process of taking a class at the university is shown in Figure 4.1. For this process, the boundaries are the point in time that the student begins the registration process through the time that grades are reported to the student at the end of the course.

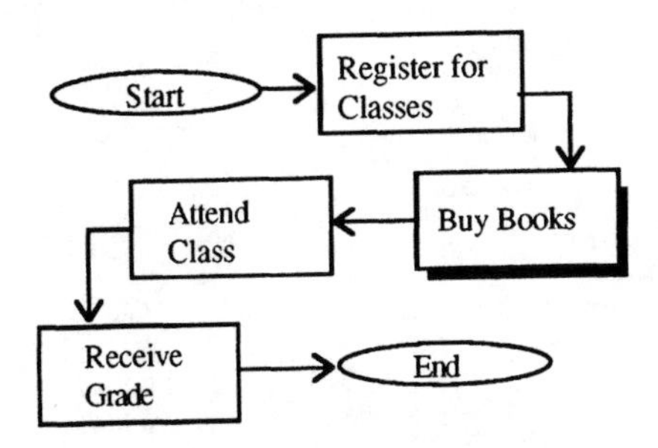

Figure 4.1 Process Flow Diagram–Taking Classes

This diagram implies that a more detailed Process Flow Diagram exists for the step listed as "Buy Books." Figure 4.2 shows a detailed Process Flow Diagram for "buying a textbook." The boundaries for the process that is shown cover one trip to the bookstore for the purchase of one book. This diagram would not serve as a complete description of the step "buy books." A complete Process Flow Diagram to explain this step would have to take multiple trips and multiple classes into account.

Another useful type of Process Flow Diagram is the deployment flow chart. In this case, a matrix is constructed with a list of the people involved on top of the page and the steps of the process shown down the page. Figure 4.3 provides an example of the grocery checkout process. The diagram shows who is responsible for each step of the process and the order in which the steps are to be performed. Shared responsibilities can be shown. For example, the stock clerk and the cashier can both bag groceries. In addition, it is possible to show how two steps of the process can be done simultaneously–by different people. In the grocery example, the stock clerk can check the price of an item while the cashier scans other items, or the stock clerk can bag groceries while other groceries are being scanned. The location of the rectangle labelled "Bag

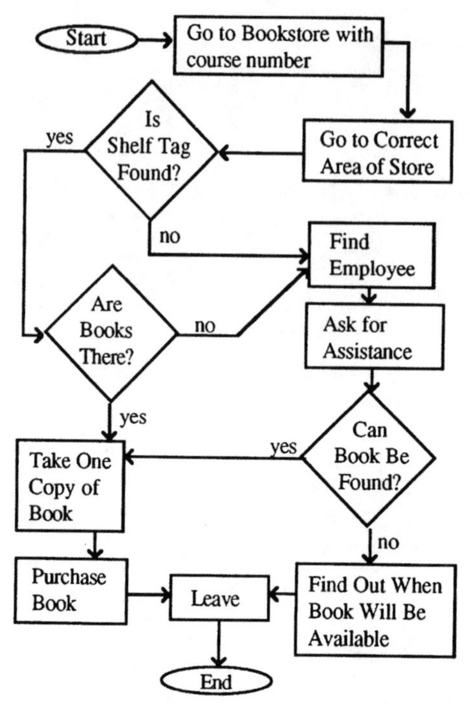

Figure 4.2 Process Flow Diagram–Buying a Textbook

Groceries" (at the bottom of the diagram) should not be interpreted as a signal that the stock clerk must wait to begin until the cashier finishes all previous activities.

Input from several people with first-hand knowledge of the process is important regardless of the type of Process Flow Diagram developed. People with different perspectives should be involved in the construction of the diagram. For example, you may want to include the suppliers to the process, the customers of the process, workers from different shifts or departments, and people who have some authority to change the operation of the process. As the diagram is developed and people from outside the group view the diagram, changes may be needed. Do not expect to draw the final product on the first try. Learning will take place as the group studies the current views about the operation of the process.

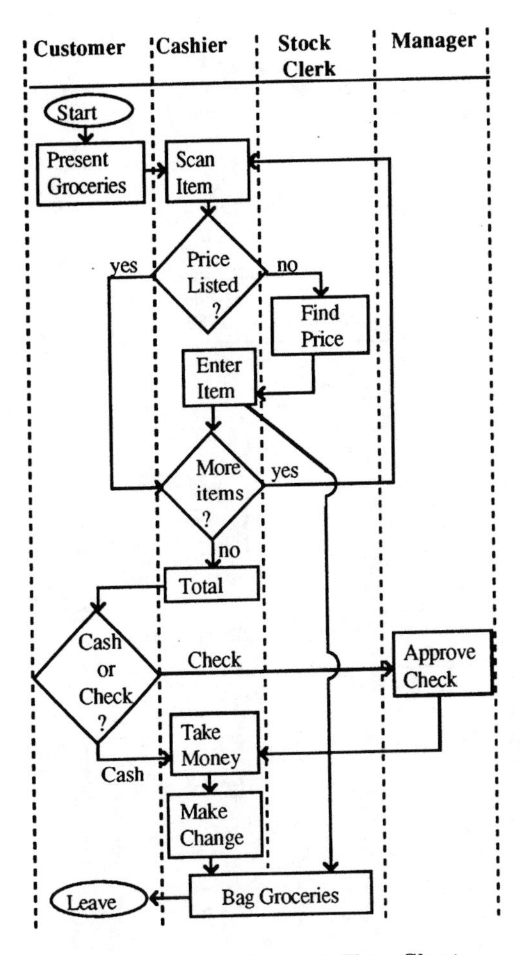

Figure 4.3 Deployment Flow Chart

Evaluation

The evaluation of Process Flow Diagrams takes two approaches. The first part of the evaluation process is determining if the diagram is complete and accurate. Once this has been done, the diagram can be used to start to analyze the process itself.

Evaluating the accuracy of the diagram requires some knowledge of the specific process being described. A Process Flow Diagram that works in one setting may not be appropriate in another setting (even though the same outcome is desired). The process described in Figure 4.4 illustrates this point. The desired outcome is the safe evacuation of a hotel during a fire. The Process Flow Diagram shown was found on the back of every door in a motel near Cincinnati, Ohio. This door opened into a hallway that ran the length of the motel. A person unfamiliar with the motel would probably think the diagram provided reasonable instructions (although taking the key to the room would not be high on most guests' lists of important things to remember).

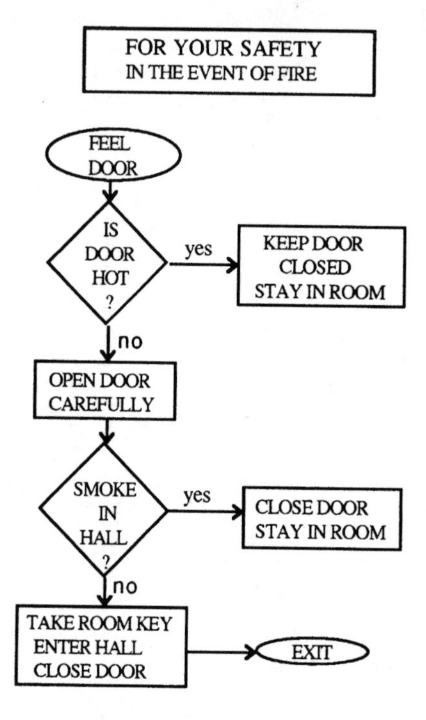

Figure 4.4 Evacuation in Case of Fire

People familiar with the motel would have a different opinion about the correctness of this Process Flow Diagram. The building has two stories. The floor of the first level is about three feet below ground level. First floor rooms have a sliding glass door that opens onto a patio with a three foot high brick wall up to the surrounding ground level. Second floor rooms have a sliding glass door that opens onto a balcony. The distance from the balcony floor to the ground below is less than five feet. The logical directions for evacuation of this motel would involve opening the sliding glass door and moving away from the building!

Often, determining the completeness of the diagram does not require knowledge of the process. For example, make sure that all blocks are connected. Also, you should be able to trace each possible path through the process from a starting oval to an ending oval. Each decision step should show what happens with each possible outcome. Figure 4.5 shows an example of an incomplete diagram. In this example, the diagram does not show what happens to assembled parts that are not good.

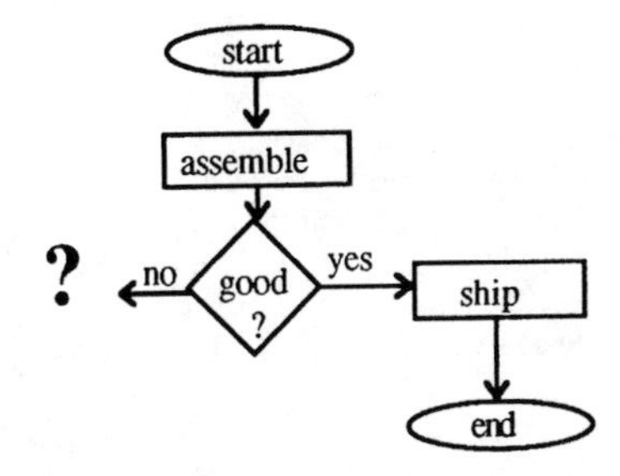

Figure 4.5 Incomplete Process Flow Diagram

With a complete Process Flow Diagram in place, you can start to use the diagram to direct efforts to learn about the operation of the process itself. The diagram can point out sources of variation.

Each decision step introduces variation into the process. What proportion of the time is each decision made? How does this affect the output from the process? How does this affect scheduling of people, ordering of materials, and sales and profits of the company? The action directed at a decision step may return the process to some earlier step or create parallel operations. For example, an inspector may pass some items to the next station, others to rework, and others to be disassembled with salvageable parts returned to inventory. The causes for these actions need to be determined.

Variability also exists in the steps denoted with boxes. Each step will have certain characteristics that are important. Use the grocery checkout example (Figure 4.3). One step calls for the cashier to scan items. The number, size, weight, and the spacing of the items on the belt will vary from one customer to another. These inputs could affect the time required to scan the items or the "attitude" of the cashier–both characteristics that are often listed as important to customers.

Cautions

Developing Process Flow Diagrams is actually a process. Teams must enter this process with a desire to understand and standardize the process. This usually means that some people will have to change the way they do things. If the team members and the people affected by the work of the team do not agree with this purpose, work to create the diagram may be a waste of time. Worse yet, it may create battles within the organization (competition between groups who see the process differently).

Development of Process Flow Diagrams that reflect the actual operation of a process requires trust. Fear will lead people to report what they think someone wants to hear rather than what is actually happening. Fear may take many forms. Some people fear being reprimanded (for not following the Standard Operating Procedure); some fear job loss (improvements may reduce the number of employees needed); some fear a cut in future income (people with a "better" way to do things may not want to share their secrets because that will allow other people to perform better–making them look better at raise time); and others fear change. The elimination of fear is not a statistical issue, but the existence of fear must be recognized in the analysis of results.

Many of the problems that can surface when a group attempts to use Process Flow Diagrams (or other tools that will result in change within an organization) involve group dynamics and psychology. A good facilitator will be able to deal with many of the issues that arise. Unfortunately, the facilitator will seldom have the ability to make the changes in organizational structure that may be needed to insure free and open expression in the group. This limitation should not be viewed as signal to accept the status quo. Good facilitators will learn to enlist non-team members in various aspects of the improvement process–thus increasing the willingness of the organization to accept change.

Next Steps

If teams stopped working on processes when they finished developing Process Flow Diagrams and started using the agreed upon current best procedure, huge improvements would usually be recognized. Of course, the use of Process Flow Diagrams should be the start–not the end–of improvement efforts.

With the identification of the steps of the process, outcomes of interest (at each step) can be identified, internal customer-supplier relationships can be identified, and direction for appropriate data collection can be developed. Once the outcomes of interest are identified, variability in these outcomes can be studied, and the sources of this variability can be explored. The use of Cause and Effect Diagrams (Module 5) can help organize your thoughts about the sources of variability.

Exercises

4.1 Draw a Process Flow Diagram for the process of using an ATM (24 hour bank machine) to withdraw cash.

Starting point: You enter the line in front of the machine.

Ending point: You receive your card and receipt at the end of the transaction.

4.2 Draw a Process Flow Diagram for the process of checking a book out of the university library. (If you have never done this, go to the library and try it!)

Starting point: You enter the door to the library with the title of a book.

Ending point: You go out the door to the library.

4.3 Draw a Process Flow Diagram for the process of buying a meal at a cafeteria.

Starting point: You enter the cafeteria.

Ending point: You sit down at a table with your food in front of you.

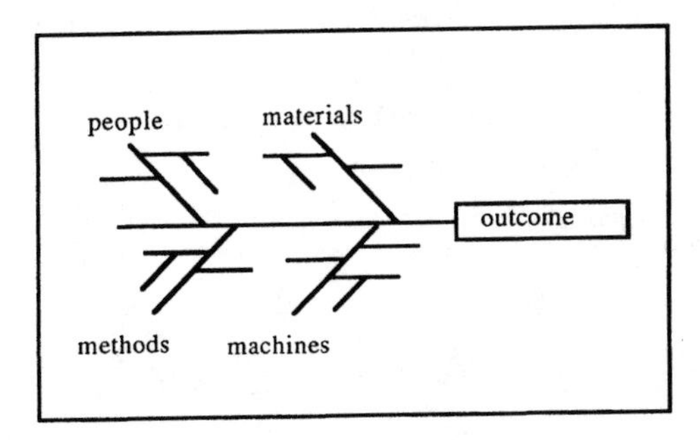

Module 5: Cause and Effect Diagrams

Cause and Effect Diagrams (also called Fishbone Diagrams or Ishikawa Diagrams) offer an organized approach to brainstorming. The completed diagram shows similar causes grouped together. The lines used to connect the causes to the effect and provide more detail about individual causes produce a figure that resembles the bones of a fish (thus the name "Fishbone"). Large branches (or bones) are explained in more detail by attaching smaller branches (or bones). If a team is used to develop the diagram, the group will develop a shared sense of the potential relationships between inputs and outputs.

Construction

The effect, output, or outcome of a process is shown on the right side of the diagram. Once an effect is identified, major causes for variability can be shown on the large "bones." Then the team can start to identify possible reasons why the effect differs from time to time. Some teams like to start with a list of possible reasons and group them later. Other teams like to take one major cause at a time and exhaust their knowledge about that cause before moving to another major cause. Still other teams like to determine the major causes of variability. Then they work freely on the details.

All of these approaches will work as long as a few "ground rules" are established.

- Initially, only common causes of variability should be shown. These are causes that affect every outcome of the process (although their effect will vary from one observation to another).

- As in any form of brainstorming, you should refrain from evaluating responses at the time they are given.

- The person giving the response should be allowed to tell where the response belongs (on which bone).

- Although some responses could be placed on more than one bone, make a choice and show it once.

- Choose major categories that make sense. The three most common approaches are:

  - "4M's" (Early definitions of a process described a process as a blending of materials, methods, machines, and men. More recently the term "men" has been replaced with "people," and environment and measurement are often listed in addition.)

  - "PEEP" (Service Organizations often use the categories of people, equipment, environment, and procedures.)

  - "Steps of the Process" (Each major step of the process can be shown as a major bone. In this case, some sources of variability may show up on multiple major bones.)

- At each step of the process, further detail is developed by asking, "What is it about this cause that results in different values for the effect?".

Consider what happens when tests are returned to students. Everyone hears the same lectures, has the same textbook, has class at the same time, and takes the same test, but the grades vary. Why? A Cause and Effect Diagram could be used to help understand why grades on an individual test vary. The effect ("Grades in a class vary") is placed in a box on the right side of the paper. Suppose that we choose to use the "4M" approach to describing major sources of variability. One of the major sources of variability would be "people." That is, grades vary because there are different people involved in the testing process. This is shown as a major bone on the diagram in Figure 5.1

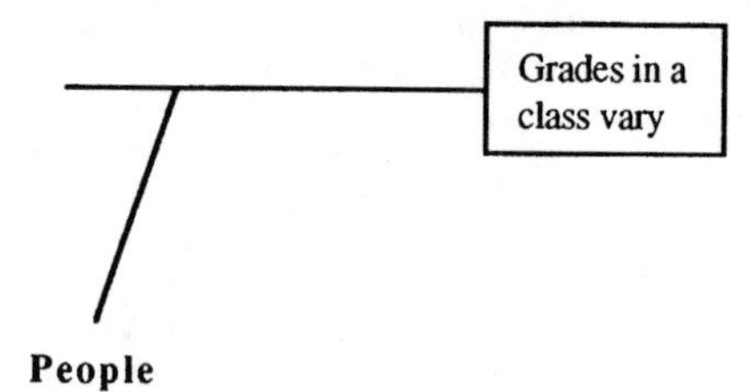

Figure 5.1 Cause and Effect–Major Bone

The next question should be "What is it about people (or which people) that causes variability in the grades?" Variability may be caused by the instructor, the student, the typist, etc. Bones for each of these would be added. Figure 5.2 shows the added detail of the student. Most people would agree that different students would be expected to perform differently on the same exam.

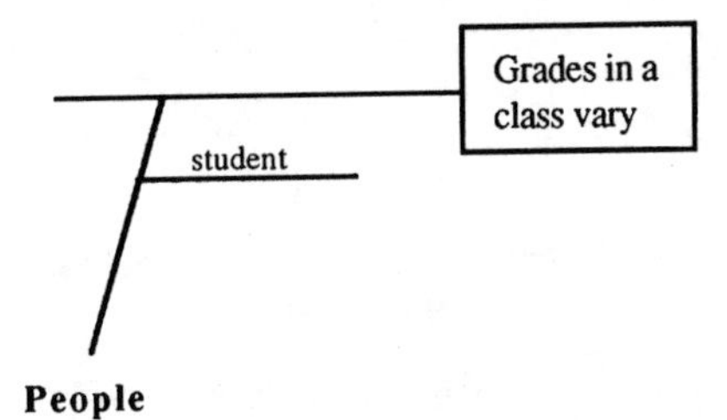

Figure 5.2 Cause and Effect–Second Level of Detail

With this added detail the questions become more detailed. "What is it about students that cause grades to vary?" Many responses are possible–interest in course, major, performance in prerequisites, health, etc. Some people may be tempted to put such things as sick or no sleep on the diagram. These do not affect every occurrence of the process, but the student's health and the amount of sleep they had the night before the exam will affect every occurrence. Figure 5.3 shows the added detail of a student's health.

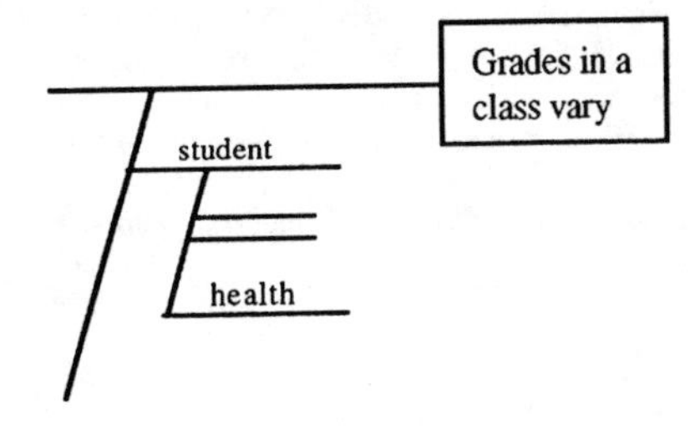

Figure 5.3 Cause and Effect–Third Level of Detail

The student's health can be further described in terms of their physical and mental health. These modifiers are shown in Figure 5.4.

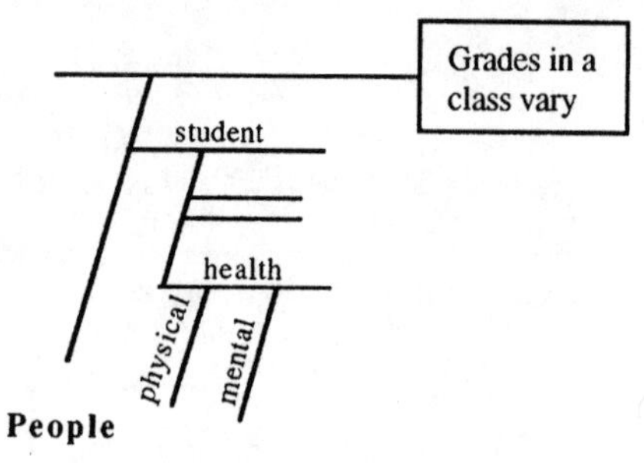

Figure 5.4 Cause and Effect–Fourth Level of Detail

Each bone and sub-bone would need to be expanded to list as many sources of variability as the team can identify. The whole group can develop one chart, or the group can be divided into smaller groups to develop individual charts that will be combined into one comprehensive diagram.

Examples

A Cause and Effect Diagram for the "Grades Varying" example is shown in Figure 5.5. The diagram shown was developed by five groups of students in an off-campus class. All students in the class held full time jobs in addition to taking one course. They were given approximately ten minutes to brainstorm and draw their group's diagram. The five diagrams were combined to produce the one that you see. With more time, more detail could have been provided.

Figure 5.6 shows another partially completed cause and effect diagram. This diagram is a combination of four diagrams constructed in a plant that makes the soft drink bottle caps described in Module 2. The plant operates four twelve hour shifts. Workers and supervisors for each shift started developing a chart. The four charts were combined to produce the diagram that you see. (A few changes were made to preserve confidential information.)

Evaluation

As a team develops a Cause and Effect Diagram they will begin to recognize where they have a lot of knowledge about the operation of the process and where they lack knowledge. A major bone may have few minor bones for three reasons. First, the team may not understand that aspect of the process. Second, the team may have developed such in-depth knowledge about that aspect of the process that the sources of variability that could show up in that area have been eliminated. And finally, some processes are people intensive and

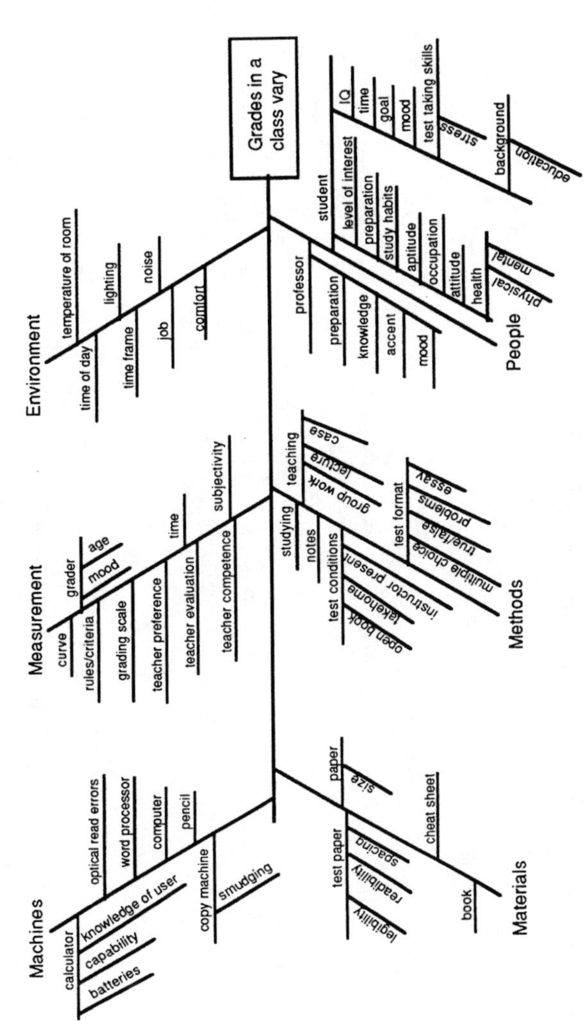

Figure 5.5 Grades in a Class Vary

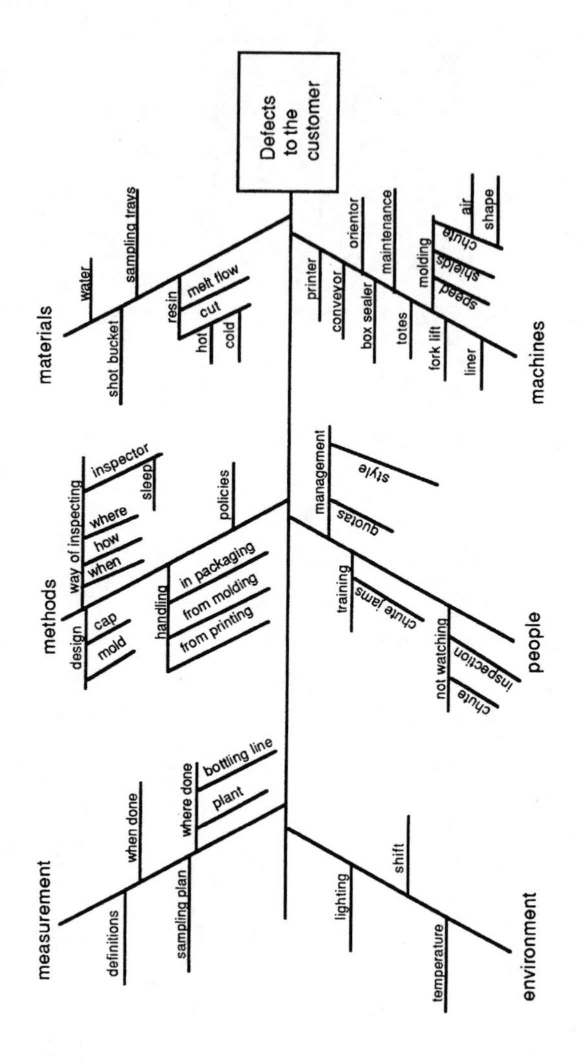

Figure 5.6 Defects to the Customer

some are not; the same is true for each major bone. The group will "get a feel" for why some bones are denser than others. This knowledge will help direct future study of the process.

Cause and Effect Diagrams that do not show multiple levels of detail (small bones on the medium bones on the large bones) usually reflect lack of knowledge about sources of variability affecting the output or outcome of interest. This could be an indication of the wrong team composition, fear of freedom of expression by team members, or lack of concentration on the topic. In any case, more common knowledge of the process needs to be developed.

Cautions

Cause and Effect Diagrams are not a stopping point in process studies. This diagram offers a way to organize thoughts and see hypothesized relationships. It does not confirm the relationships or give any documentation to support the strength of the relationships.

Effective use of Cause and Effect Diagrams, like Process Flow Diagrams, depends on people's willingness to express themselves freely. Sometimes people will begin to share ideas if you precede the Cause and Effect Diagram with a variation referred to as a Negative Ishikawa. As the name implies, participants are asked to develop a diagram showing what could go wrong. The effect is still listed to the right, but this time the diagram is filled with all of the special causes people can describe. Sharing "war stories" will break down some of the barriers between people and allow them to be more open in other discussion.

Next Steps

Cause and Effect Diagrams identify possible relationships. They do not provide proof that the suggested cause and effect relationship is appropriate. Variables for data collection can be identified, and a plan for data collection can be developed based on the team's increased knowledge of the process. The collection and analysis of data will confirm (or deny) the existence of relationships. Where relationships are found, the data will help determine the strength of these relationships.

Even the most elementary Cause and Effect Diagram should help people recognize that no one cause is individually or independently responsible for the variability in the output. Understanding how these inputs combine is the key to understanding, reducing, and controlling the variability in the output.

Exercises

5.1 Draw a Cause and Effect Diagram describing why "the amount of learning" varies in a class.

5.2 Draw a Cause and Effect Diagram to describe why the time to travel to school varies from day to day.

5.3 Draw a Cause and Effect Diagram explaining why a person's phone bill varies from month to month.

5.4 Draw a Cause and Effect Diagram to help understand why a car's mileage (mpg) varies from tank to tank.

5.5 Draw a Cause and Effect Diagram showing why the number of students attending class varies from day to day (same class).

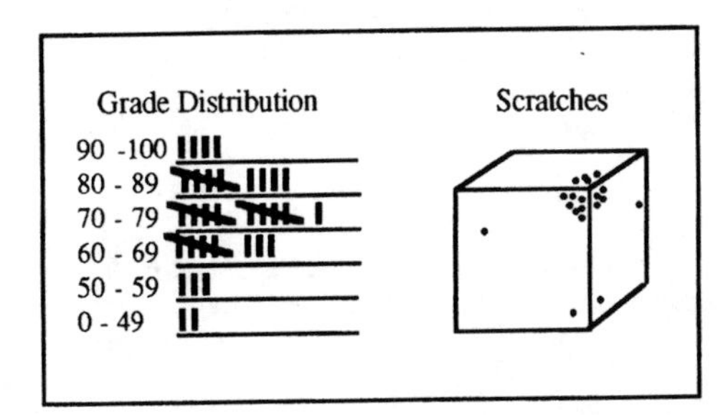

Module 6: Check Sheets

Check Sheets provide a way to simplify data collection. They provide data collectors a standard format for reporting observations. This standard format helps reduce variability due to the method of recording the data and prompts all collectors to provide answers to the same questions. Check Sheets do not have one set format. In fact, "check sheet" can signify any form that is developed to help collect data.

Construction

Before any Check Sheet can be developed, the team must identify what characteristics of the process are to be observed and operationally define these characteristics. Variables that may affect the values observed for the characteristics of interest should be identified. For example, the color of a box may be of interest. The team may think that the shift or manufacturing line that produced the box may affect the color (since each line and each shift does their own set up). The Check Sheet that is developed should account for these possible sources of variability. This can be done in two ways. Data can be collected and analyzed separately from each line and shift. Or, data can be collected from some point that uses output from all lines and shifts. If this approach is taken, the Check Sheet should provide a place to record the line and shift that produced each observation. Analysis of this data should consider the possibility that the line or shift will affect the color observed.

Check Sheets can be graphical or tabular. The graphical version is referred to as a "Concentration Diagram" or a "Defect Concentration Diagram." This form of a Check Sheet shows a drawing of the product that is being observed. The same sheet is used as multiple items are checked. Defects found on the item are marked on the drawing. After multiple items have been checked, areas that are repeatedly causing problems will be identified.

The tabular form of a Check Sheet should be developed with consideration for the person that will be doing the collection. Terms should be clearly defined. Writing should be legible. Layout should be logical. Space should be adequate for recording responses.

Examples

The simplest form of the Check Sheet is a Concentration Diagram. Figure 6.1 shows a concentration diagram used by a company that prints cigarette cartons. The printing process operates in such a way that each revolution of the print cylinder creates four prints. Each print position is inspected separately. The concentration diagram shown here represents output from one position. The dotted lines represent scoring (the indentation that makes folding the box easier) on the carton.

Cigarette Carton

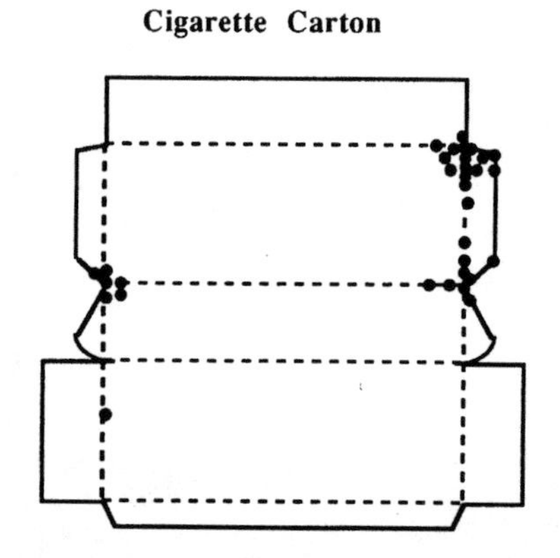

Figure 6.1 Defect Concentration Diagram

One quick glance indicates problems with the scoring along one portion of the carton and three corner cuts. The one problem on the lower left does not appear to be a consistent problem–at least not at the same level as the other problems identified.

Tabular Check Sheets may be used to collect data on one or more variables. Figure 6.2 shows a Check Sheet that could be used to determine the current use of word processors in an organization that uses IBM (or Compatible) and Macintosh computers. This would be considered a Check

Sheet for one variable (You could say, let X = name of word processor used. The possible values for X would be "Word Perfect," "Word," "Wordstar," etc.). You may want to include separate lines for different versions of software (if the people responding to the question would know the version).

WORD PROCESSORS

IBM or Compatibles

Word Perfect ____________
Word ____________
Wordstar ____________
Works ____________
WriteNow ____________

Macintosh

Word ____________
MacWrite ____________
Word Perfect ____________
Works ____________
WriteNow ____________

Figure 6.2 Check Sheet for Word Processor

The word processors are listed in an order that represents the expected proportion of responses (for ease of the data collector). For example, Word Perfect seems to the most popular word processor in the IBM environment, and Word appears to be the most popular in the Mac environment (Spring 1992). In addition to the information listed, you may want to provide a place for the date and lines to list other word processors.

As the number of variables increases, the development of the form requires more thought and care. More thought in the development of the Check Sheet may provide more consistent results (in terms of the types of measurements reported). Figure 6.3 shows a Check Sheet for collecting data about "traffic flow" through a specific intersection. Of course, data collectors would need to be trained to use this form. Part of the training would include making sure that everyone has the same definitions of "through the intersection," "pedestrian," "car," "truck," and "bus." Collectors would need to know how to classify motorcycles, bikes, roller blades, and skateboards.

The Check Sheet has a built in diagram of the area that is being observed. The location of the data collector is shown on the diagram. Arrows show the direction of traffic flow (simplifying the job of the data collector). Since motorized vehicles cannot travel diagonally through the intersection, these blocks are shaded. There is a place to note the time covered by the form. Since this intersection is located in the heart of Virginia Commonwealth University

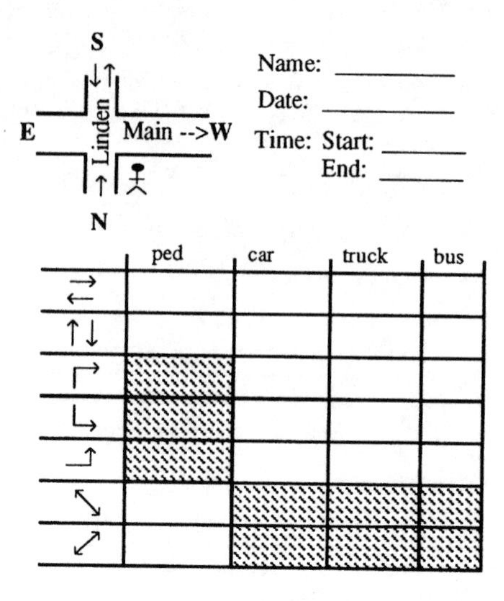

Figure 6.3 Check Sheet Using Visual Guides

(a couple of miles west of downtown Richmond), the time of day and day of the week would be expected to be an indicator of traffic flow. For example, rush hour traffic will use Main (westbound) in the afternoon, and pedestrians will use the intersection between classes. Before this Check Sheet is used for collection of data, the collectors should conduct a "pilot" study. The results of this study will provide input to improving the form before it is used on a large scale.

Evaluation

Check Sheets provide data as input to a variety of tools for analyzing processes. Periodically, Check Sheet should be studied to confirm that the data collection procedure is consistent. If you find some common thread (such as one worker, one day of the week, or one time of day) that leads to observations that differ from the others, consider the need for clarification or additional information to be collected.

Cautions

Good Check Sheets are not easy to design. Time spent in this phase of a process study will be well spent. The best statistical techniques cannot compensate for poor data collection. Terms must be operationally defined, problems in data collection must be anticipated, and workers must be trained.

You should not assume that a process that involves different days, different shifts, or materials from different suppliers operates similarly under these different conditions until you have data to support that conclusion. Check Sheets should provide space to record values of variables that are potential causes of variation in the output.

Next Steps

Check Steps are an intermediate step in process studies. Sources of variation must be identified (possibly through Process Flow Diagrams and Cause and Effect Diagrams) before Check Sheets can be designed. The type of analysis planned should be taken into account when the specific content of the Check Sheet is determined. The team should ask themselves, "What data would I need to make a decision?" The data collected on the Check Sheets will be used to analyze the process with such tools as Pareto Charts, Run Charts, Control Charts, Histograms, and Scatter Diagrams.

Exercises

6.1 Develop a Check Sheet that could be used to record the number of cars from various manufacturers parked in a parking deck.

6.2 Develop a Check Sheet that could be used to tally the number of people wearing various types of athletic shoes.

6.3 Develop a Concentration Diagram that could be used to study the level of use for desks in a classroom with fixed position seating (desks that are bolted to the floor).

6.4 Some people claim that students who sit in the front of class make better grades. How could a concentration diagram be used to assess this claim?

6.5 Develop a Cheek Sheet for recording the major of students in this class.

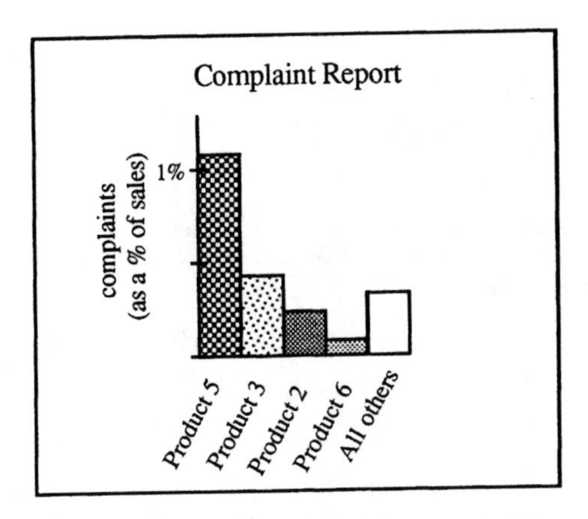

Module 7: Pareto Charts

Pareto Charts are used to show the relative importance of qualitative or categorical data. Displaying the categories in order of magnitude, from largest to smallest can be used to make decisions about where to direct resources. Categories are often ranked based on frequency of observation, relative frequency of observation (proportion/percent), or on some basis of cost. Typical Pareto Charts will show that most problems (or occurrences) come from a small number of categories.

Construction

The first step in the construction of a Pareto Chart is to determine the categories of interest. Next, data must be collected to determine the frequency of occurrence for each category (Check Sheets can help). Suppose you are assigned to develop a program for the local Girl Scout Council. Further, suppose that you do not have the time, money, or help to develop more than one program. Where should you concentrate your efforts if you want to serve the most girls. You are told that more than 11,000 girls are served in five age levels (grades levels): Daisy (K-1), Brownie (1-3), Junior (3-6), Cadette (6-9), and Senior (9-12). The local council provides data in a form that makes sense to them–from youngest girls to oldest girls. The percentages reported are shown on the following page.

Age Level	Percent of Membership
Daisy	5.73
Brownie	46.05
Junior	31.96
Cadette	8.18
Senior	1.95
Other	6.13

The next step in making a Pareto Chart requires you to rank the categories (age levels in this case). If a "catch all" category of "other" is used, list it last. The result of this step is shown below.

Age Level	Percent of Membership
Brownie	46.05
Junior	31.96
Cadette	8.18
Daisy	5.73
Senior	1.95
Other	6.13

The Pareto Chart shows a scale on the vertical axis. Depending on the situation, this scale may represent a percent, a number of occurrences, the cost of occurrences, or some other measure. The categories are shown on the horizontal axis in the same order that you ranked them. Bars are placed above the categories. The height of each bar is determined by the characteristic that you ranked (in this case, percent of membership). A title is placed at the top of the chart. A Pareto Chart for the Girl Scout Membership example is shown in Figure 7.1.

Girl Scout Membership

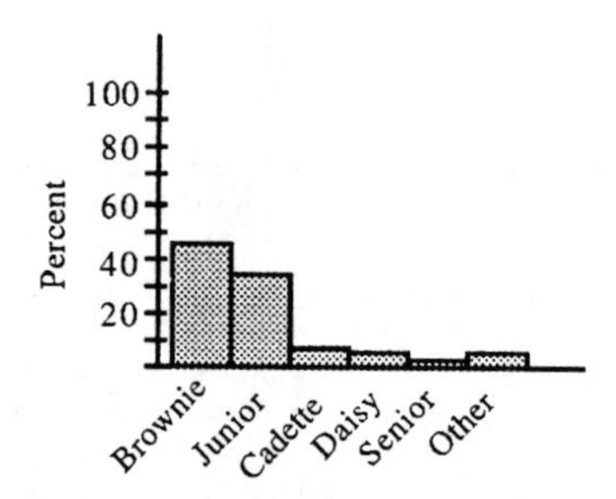

Figure 7.1 Membership by Age Group

The numbers told the same story that the graph shows, but most people comprehend information from pictures faster and clearer than things presented in words or numbers alone. Clearly, both presentations of this data indicate that Brownies and Juniors make up most of the membership of this council.

Some people add a bar showing the cumulative effects. In order to do this, start with the ranked list from the original Pareto Chart. For each category add the measure for that category to measures for all categories listed above it. Cumulative percents for the Girl Scout example are shown below.

Age Level	Percent	Cumulative %
Brownie	46.05	46.05
Junior	31.96	78.01
Cadette	8.18	86.19
Daisy	5.73	91.92
Senior	1.95	93.87
Other	6.13	100.00

For each category locate a point on the Pareto Chart directly above the right edge of the bar at a height given by the cumulative value given for that category. For example, you would place a point above the right side of the bar for Seniors at a height of 93.87. Once you have placed all of the points, draw line segments to connect them. Figure 7.2 shows the result for the Girl Scout example.

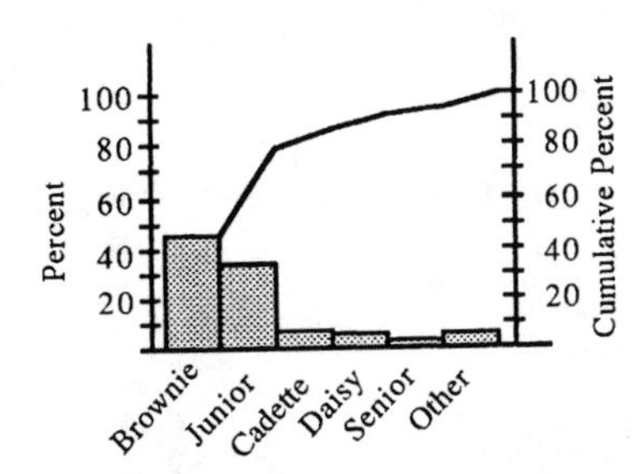

Figure 7.2 Membership–With Cumulative Percent Shown

For this problem, the value of the cumulative line for a specific category can be interpreted as the proportion of membership you would have served if you planned a program for this specific age level and all age levels lists before it. For example, the cumulative percent for Cadettes is 86.19. Brownies and Juniors are listed before Cadettes. If you planned a program that reached all

Cadettes, Juniors, and Brownies, you would have reached 86.19% of the membership.

Examples

Many times Pareto Charts can be expanded. For example, from the Girl Scout example you may have decided to plan a program for Brownies. You could further refine the type of program you are planning by looking at the Brownie membership. Are these girls from cities or rural areas? Do they live in the city, north of the city, south, east, or west? Two other Pareto Charts follow. Figure 7.3 shows the Brownie membership by type of community. Figure 7.4 shows Brownie membership by location.

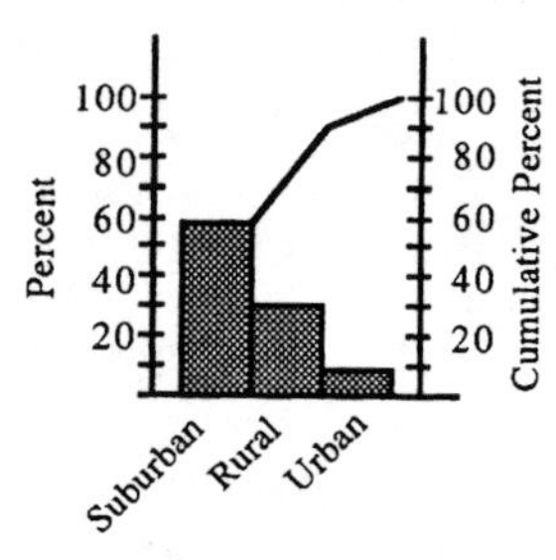

Figure 7.3 Brownie Membership by Type Community

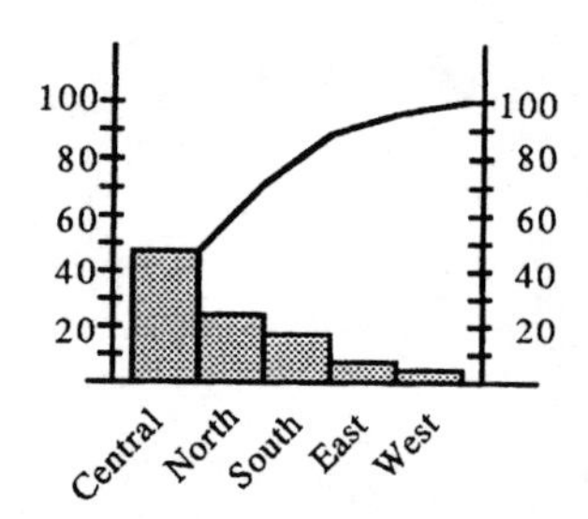

Figure 7.4 Brownie Membership by Location

Based on this information, you may decide to plan an activity for Brownies in a central location that builds on the assumption that most of the participants are living in suburban areas. From a more global perspective, these same Pareto Charts could be used to help determine the number and type events to be planned for different areas of the council.

The Pareto Chart shown in the box at the beginning of this module provides a more typical use of a Pareto Chart. Notice that the vertical scale is given as complaints (as a % of sales). "Number of complaints" would not be an acceptable choice for the vertical axis unless equal quantities of each product were sold. Suppose you knew that there were 10 complaints about Product 3 and 5 complaints about Product 2. Does this indicate more problems with Product 3? Not necessarily. How many of each product were sold–one million of product 3 and 6 of Product 2!

Evaluation

Pareto Charts provide a quick graphical evaluation about the relative importance of the categories studied. Often, changing the view, by changing the characteristic on the vertical axis, will result in different rankings. Careful choice of the characteristic is important. Try different vertical axes and different category choices to get a better view of what is happening. Some choices for vertical axis include cost, number of occurrences, and percent of occurrences. Categories may be such things as type problem, shift, machine, location, or worker.

A Pareto Chart developed before an improvement effort can be compared to another Pareto Chart developed after changes have been made. If the improvement effort has been effective, the results should be predictable. The category selected for improvement should show a reduced level of occurrence and possibly a change in relative rank.

Cautions

Any time categories are ranked, one will be listed at the top! This sounds trivial, but we often overlook this obvious outcome. The fact that one category ranks highest does not guarantee that something different affected that category. Recall the "a's" that you made in Module 2 (Example 3). You were asked to pick the "a" you liked best and determine what you did differently when you made that "a." There was not one specific thing that caused that "a" to differ from the other ones–it was just the chance combination of a lot of factors that affected all of the "a's." Asking "Why is this one different?" was not appropriate.

Take another example to help understand some of the problems with "jumping" on the category that ranks highest. Toss a pair of (fair) dice 60 times. Make a Check Sheet to record the number of dots appearing on the top face each time. You should have 120 observations when you finish. If the dice are fair, you would expect to see each face the same number of times–but it does not happen. Does this mean that the face that was seen most often in these 60 tosses of the dice will be seen most often in the next 60 tosses? If you answer no–and if the dice are fair, you should say no–then you should not use the ranking to direct action. On the other hand, if you find that the relative ranking of categories remains the same over time, you have justification to use the ranking to direct action.

Often the categories chosen are not independent. Changing something in the process, in an attempt to reduce (or increase) the number of occurrences of one outcome, will usually affect other categories. In our Girl Scout example, the success of programs planned for this year's Brownies will affect next year's Brownie and Junior membership figures (since girls will decide whether to return to scouts based on experiences the previous year).

Next Steps

We must be able to determine when to use the ranking of categories to direct action and what types of actions are appropriate. Both of these issues deal with understanding how to distinguish between common cause and special cause variability. Run Charts and Control Charts are the tools we use to make this distinction.

Exercises

7.1 An auto insurance company has experienced a problem with incomplete applications arriving for processing. In the past week 1200 applications were incomplete. Some applications were missing more than one entry. The following omissions were observed:

Type Omission	Number
Year of Car	90
Driving Record	700
Model/Make of Car	60
Date of birth of Applicant	150
Serial Number of Car	300
Address of Applicant	50
Other	75

Draw a Pareto Chart showing the types of omissions.

7.2 Draw a Pareto Chart to display the number and types of Master's degrees awarded by a large university in May 1992.

Degree	Number
Master of Arts	31
Master of Fine Arts	49
Master of Science	253
Master of Health Adm.	28
Master of Business Adm.	74
Master of Public Adm.	42
Master of Education	108
Master of Nursing	53
Master of Social Work	191
Other	53

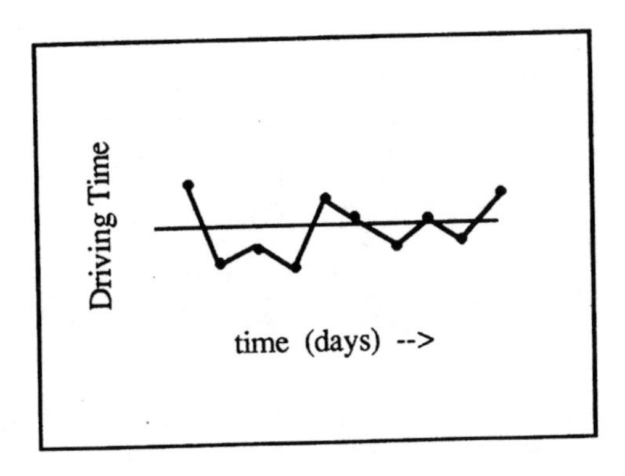

Module 8: Run Charts

Run Charts are used to visually represent the way some characteristic of a process changes over time. Run Charts indicate if the process average is changing over time. Trends and sustained shifts in some measured characteristic will show up as patterns in points plotted on a Run Chart. When a process is stable, plotted points should vary about some average value; about half of the points should be above the average and about half below the average; and no pattern should describe which points will plot above the average and which will plot below the average (i.e., you should not expect the first half of the points to be above average and the last half below, or you should not expect every other point to be above the average). Run Charts **do not** indicate how much spread you should expect to see in the points.

Run Charts can be used to track many different types of information. Run Charts can be used to plot proportions, number of occurrences, measurements, averages of groups of measurements, or the spread in groups of measurements. In fact, any characteristic that is repeatedly reported and used for planning purposes should be plotted on a Run Chart.

Construction

The construction of Run Charts is easy. The horizontal axis represents time or some other sequence (such as order of production). The vertical axis is scaled to show possible values for the characteristic being tracked. Points are plotted in the order they were produced. Connecting the points allows a "motion picture" to develop. Both axes are labelled (so information is conveyed to someone other than the person making the chart), and a title is placed on the chart.

The choice of the scale for the vertical axis depends on the situation. The scale should be chosen so that the variability that exists in the process can be seen. If the measurements will range from .998 to 1.003 inches, labelling the vertical axis with 0 inches, 1 inch, 2 inches, etc. would **not** allow the variability to be seen. In this case, it would be more appropriate to start the scale slightly below .998 (say at .995) and continue up to slightly above 1.003 (say to 1.005). Each space on the vertical axis might represent .001 inches. But, the scale should not show more detail than provided by the measurement device. If measurements are recorded to the nearest inch, showing thousandths of an inch on the scale would not be useful.

After an initial Run Chart (describing past data) is drawn, additional points can be added as the process continues to produce output. For this reason, the initial chart should be placed on the paper in a position that will allow for additional points to be plotted.

Examples

In this section, Run Charts for five different "types" of data will be presented. Further study of each type of data will require different formulas. The first two types of data are classified as **attributes data**. Attributes data may be binary (observe a collection of items and decide if each item belongs in a given category or not; then record the number belonging or the proportion belonging), or attributes data may represent a count (count the number of occurrences of some event). The last three Run Charts will represent **variables data**. Variables data deals with quantitative data or measurements. In some cases individual data points are plotted over time. In cases where large amounts of product (or service) are produced, several items may be inspected periodically. This set of items is referred to as a subgroup. Run Charts can track the average measurement or the amount of spread in the measurements from subgroups. Using a Run Chart that tracks the average along with one that tracks the spread provides powerful insight into the operation of a process.

Example 1: One characteristic of interest in the manufacture of executive desks is the finish of the top. The desk top should be free of scratches. Each day 30 desks from that day's production are inspected. The total number of scratches found in the 30 desks is recorded. Recognize that it is possible for no scratches to be found, 1, 2, 3, or a very large number of scratches (many more than 30), but it is not possible to observe 1.5 scratches. The Run Chart showing the first 20 days is given in Figure 8.1. A "center line" showing the average for those 20 days could be added. The height of this line is determined by finding the total number of scratches recorded in all 20 days and dividing that total by 20. The addition of the "center line" helps identify sustained shifts in the process.

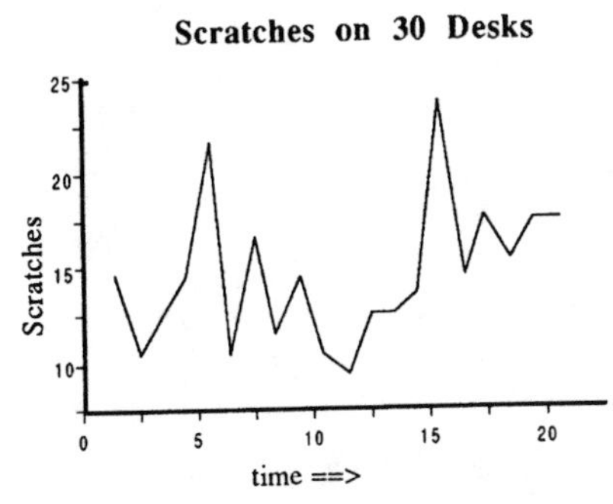

Figure 8.1 Run Chart for Scratches on 20 Groups of 30 Desks

As we should expect, the number of scratches varies from one day to the next. The Run Chart **does not** indicate the amount of variability that should be expected. Control limits (for a "c chart") are needed to determine if the highs and lows should be considered "special" or if they could be expected from this process. Control limits will be discussed in the next module.

Example 2: Grading can be viewed as a process. At the end of each semester instructors submit grades for students in their classes. Every student receives a grade of A, B, C, D, or F. Each letter grade's proportion of the total assigned grades would be expected to vary from class to class and semester to semester. One instructor decided to plot the proportion of A's submitted in required quantitative courses. Note that each student receiving a grade receives an "A" or some other grade (i.e., we could say that there are two possible outcomes for each student). The proportion "A's" for a class would be computed by dividing the number of students receiving an "A" by the total number of students receiving grades in that class. The Run Chart is shown in Figure 8.2.

This Run Chart uses different symbols to consider another potential source of variability–the course. Patterns in the symbols could be used to determine if the instructor's grading process differs from one course to another. Of course, there could be multiple reasons for consistently different outcomes for different courses. For example, the instructor may enjoy one course better than another (resulting in more interesting lectures and better grades). Another possibility is that the students taking one course may be "better" than the students taking another course (admission to Junior level courses may require a minimum grade point average but Freshman level courses do not). In this example, Course 1 and Course 2 are usually taken by students during their

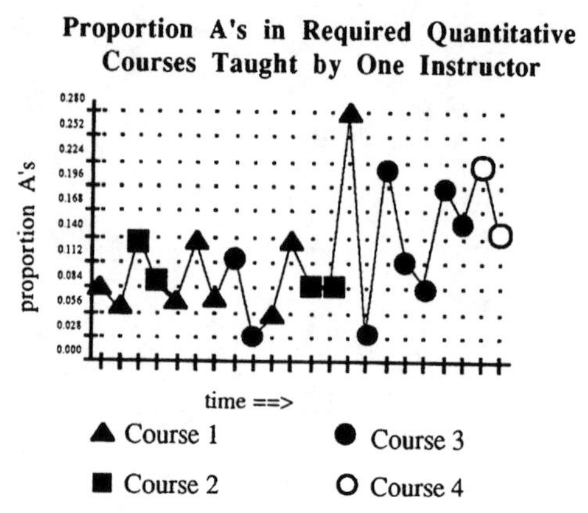

Figure 8.2 Run Chart for Proportion A's in a Class

Freshman or Sophomore year. Course 3 and Course 4 are usually taken by students who have met course and grade point average requirements to be admitted to a business degree program.

Any further analysis of the type of data shown in this example, proportions, would use control limits established for a "p chart." If we had plotted the number of students receiving "A's" instead of the proportion of "A's," an "np chart" could be used (assuming all classes were the same size). Control limits will be discussed in the next module.

Example 3: Often we receive values for the "same" characteristic over and over again. For example, we get an electric bill, a phone bill, and a water bill every month. A company reports sales, profits, expenses, and inventory on a regular basis. These numbers should be viewed as part of a sequence. Figure 8.3 shows the monthly phone bill for one household.

At first glance, the point from April 1990 appears to stand out. Should a phone bill this high be questioned (as something special), or can this household expect to see bills this large occasionally? Control limits for an "X chart" (or "individuals chart") would be needed to answer this question.

Run Charts for individual values and the associated Control Charts have applications in board rooms, managers' offices, in service organizations, and on plant floors. These charts provide decision makers a rational basis for planning.

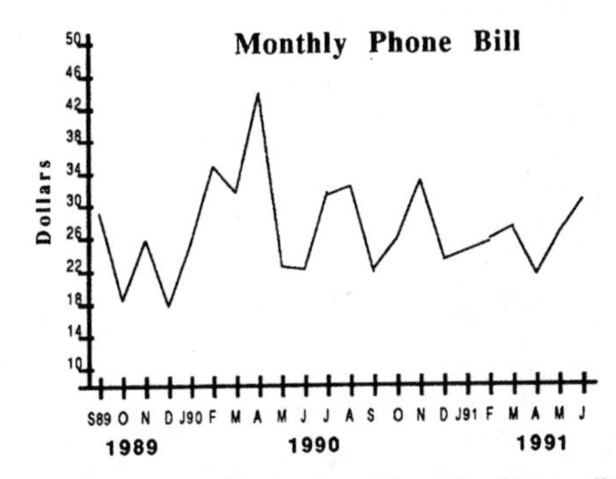

Figure 8.3 Run Chart for Monthly Phone Bill

Example 4: In a manufacturing environment, Run Charts (and the associated Control Charts) plotting the average and spread in subgroups are popular. Periodically, several items (a subgroup) are selected from the manufacturing line, some characteristic of each item is measured, the average and range are calculated, and these values are plotted on their respective charts. Consider the following example. Suppose you work in a plant that fills "5 pound" bags of sugar. Each hour you collect 6 consecutive bags from the filling line and weigh them. The weights of the bags from your most recent hour's subgroup are (in pounds):

5.035	5.04	5.075
5.025	5.05	5.065

The average of these values is referred to as $\overline{X}$ (pronounced "X Bar"). $\overline{X}$ is calculated by adding up the weights from the six bags of sugar and dividing the total by 6 (the number of bags of sugar weighed). Mathematically, the formula is:

$$\overline{X} = \frac{\sum_{i=1}^{n} x_i}{n}$$

where "i" represents the observation number, "n" represents the number of observations, and the Σ tells us to add the observations. For this example $\overline{X}$ is calculated as:

$$\frac{5.035 + 5.04 + 5.075 + 5.025 + 5.05 + 5.065}{6}$$

or $\bar{x} = \frac{30.29}{6} = 5.0483$ (the average).

The range is referred to as "R." To calculate R, find the largest value and the smallest value. Then, subtract the smaller number from the larger. The result is the range. For this example,

$$R = 5.075 - 5.025 = 0.05.$$

R provides a measure of how much the numbers differ from each other. Notice that all but two observations are ignored in the calculation. When there are only a few observations in a subgroup, this causes few problems. With large numbers of observations a different measure describing the spread in the observations is desired (one that uses all of the values). Once the subgroup size exceeds 8 or 9, the standard deviation (introduced in your textbook) is often used.

If a process consistently produces product (or service) at approximately the same average value and the same amount of spread about that average, planning for the future output of that process is simplified. Charts tracking the average and spread must be used together to determine if these conditions are met.

Example 5: A manufacturing process that makes precision molded products uses subgroups of size 5 collected hourly to monitor a certain characteristic of the product. Measurements are made to the nearest 0.001 inch. Rather than record the actual measurement, the deviation from target is recorded. For example, it the nominal (target) value is 10.000 and the item measures 10.002, then 0.002 is recorded; likewise, if the item measures 9.997, –0.003 is recorded. The Run Chart for $\bar{X}$ is shown in Figure 8.4. The Run Chart for R is shown in Figure 8.5.

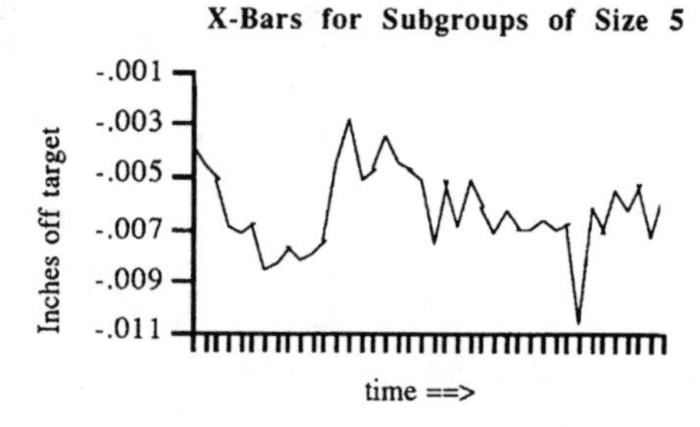

Figure 8.4 Run Chart for Subgroup Averages

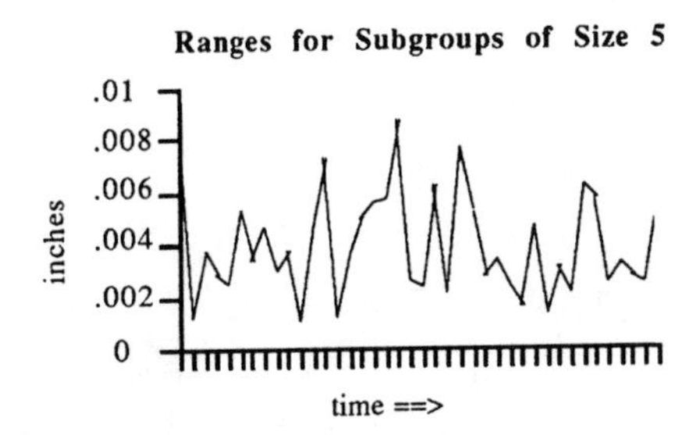

Figure 8.5 Run Chart for Subgroup Ranges

The Run Chart for the $\overline{X}$'s shows that this process is producing a product that, on average, is smaller than desired. Remember that the values plotted are averages. Five numbers were averaged to obtain each dot. Part of the individual measurements will be larger than the average, and part of the measurements will be smaller than the average. By considering the range (less than 0.008) along with the information about the average, it is safe to say that most product produced by this process will be smaller than desired.

Control limits for both the R chart and $\overline{X}$ chart would be needed before more specific comments could be made about the operation of this process. The next module will describe how these two charts will work together to provide information for improving the process.

Evaluation

Run Charts allow us to capture the information that is contained in the time ordering of the data. Specifically, Run Charts help us recognize trends and other patterns in the data.

Figure 8.6 shows a clear upward trend in the closing price of West Texas Intermediate Crude Oil during the Summer of 1990. (Iraq invaded Kuwait on August 2, 1990.) As a rule of thumb, at least seven successive increases or at least seven successive decreases can be considered a trend. The cause of the trend should be investigated.

Other patterns can be seen on Run Charts. If more than seven points in a row plot on the same side of the center line (calculated from data collected from the process), the cause for the sustained shift should be investigated. Any

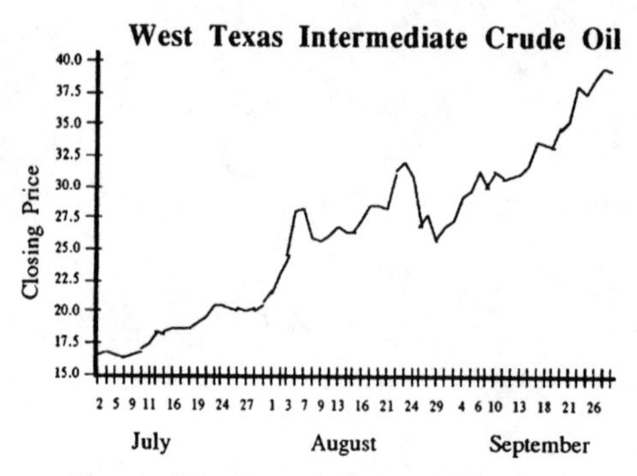

Figure 8.6 Run Chart for Oil Prices

pattern that repeats over and over again (in a predictable fashion) should be a signal for special attention. Consider the Run Chart shown in Figure 8.7.

Absenteeism has been plotted for several weeks. Notice that more people are absent on Mondays than any other day of the week. In fact, the ranking is the same week after week–Monday, Friday, Wednesday, Tuesday, and Thursday. (Thursday is payday!) Unless different symbols are used for each day, this pattern could go unnoticed.

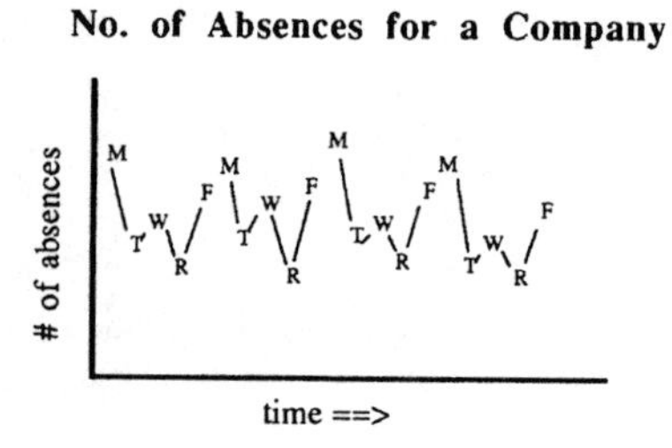

Figure 8.7 Run Chart for Absences

Cautions

Run Charts do not provide a method for determining how much variability should be expected from a process. Individual points should not be singled out and investigated. Control limits are needed to determine if the magnitude of differences is due to something other than causes common to the process.

Run Charts are great tools for visually displaying data to recognize trends and pattern in time ordered data. Not all data is time ordered (or ordered in any way that has meaning). Do not look for trends in data unless the points are plotted in some way that order has meaning.

Next Steps

Every example given in this module has referred to the need to determine how much variability is too much to have been produced by a process operating in a stable manner. Repeatedly, you have been told that control limits would be needed to further analyze the situation. It should come as no surprise that the next step is to assess process stability through the use of Control Charts.

Exercises

8.1 The miles per gallon have been calculated for one car based on highway (not in town) driving. Plot the numbers on a Run Chart and comment on any patterns. (The observations are recorded in columns.)

36.2	38.6	35.1
34.0	34.4	34.3
35.2	34.9	36.1
35.6	32.5	35.2
34.9	33.7	34.1
38.0	38.6	35.7
38.3	35.4	37.2
37.1	34.3	34.7

8.2 Some professors take attendance in class. The following numbers represent the proportion of students missing class. There are 50 students registered for the class. The first test was given during the tenth class meeting. Plot these values on a Run Chart and comment. (Observations are listed in columns.)

0.08	0.06	0.14	0.08
0.02	0.04	0.12	0.16
0.06	0.06	0.12	0.14
0.02	0.00	0.08	0.12
0.04	0.24	0.12	0.14
0.12	0.16	0.10	0.20

8.3 Applications for auto loans often contain omissions. Rather than classify each application as complete or incomplete, the number of omissions are counted on the incomplete applications. A recent study of 24 incomplete applications resulted in the following data. Plot these observations on a Run Chart and comment. (Observations are listed in columns.)

2	1	1	1
1	3	2	3
2	2	2	2
2	1	3	2
3	6	2	1
2	1	2	2

8.4 The following numbers represent one person's time (in minutes) required to travel to work. Make a Run Chart and comment. (Observations are recorded in columns.)

31	30	30	30
27	29	31	33
30	30	30	29
30	31	29	30
32	29	28	28
27	33	32	29
28	28	27	32

8.5 Draw a Run Chart showing the variation in water and sewerage bills received for one household. Each bill covers approximately 60 days. Comment on any patterns.

Month/Year	Bill
1/89	17.00
3/89	17.00
5/89	20.83
7/89	23.02
9/89	22.51
11/89	24.09
1/90	24.09
3/90	27.53
5/90	31.59
7/90	22.64
9/90	27.20
11/90	27.30
1/91	30.59
3/91	37.37
5/91	38.28
7/91	27.30

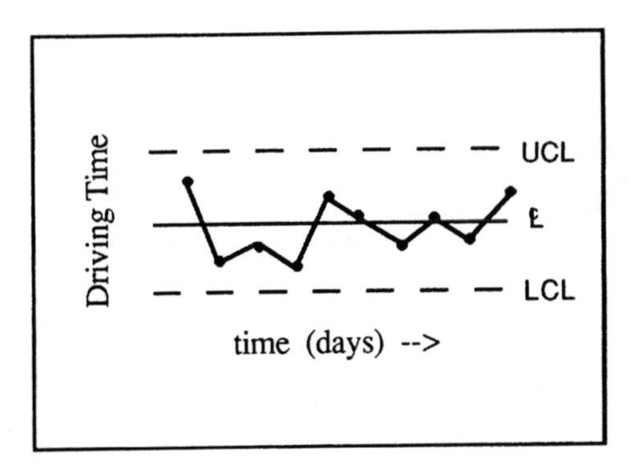

Module 9: Control Charts

Control Charts provide an operational definition for a stable process–they provide a method for viewing a process and criteria for reaching a decision. The method involves plotting points in time order (making a Run Chart) and using the **data collected from the process** to calculate limits to describe the amount of variability that should be expected from the process. Figure 9.1 shows the general appearance of a Control Chart. The charts have a center line (denoted by ℄), a lower control limit (denoted by LCL), and an upper control limit (denoted by UCL).

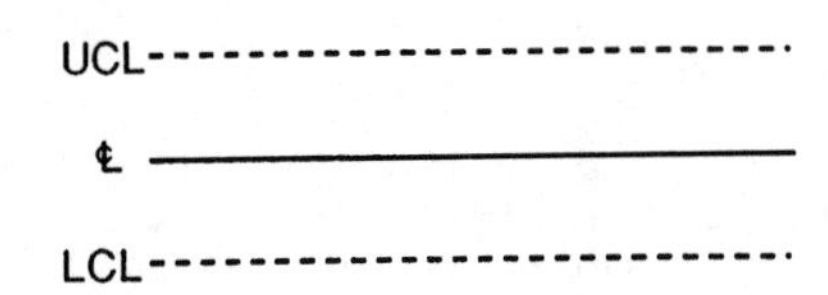

Figure 9.1 Limits for a Control Chart

The criteria for decision making involves checking to see if all points plot within the limits calculated and looking for patterns in the data. Stable processes, processes that are operating in a predictable fashion, will produce Control Charts with points plotting within the control limits and forming no patterns. To say that a process is stable (or in statistical control) says **nothing** about whether the process is producing acceptable output. A process can be stable and producing 100% unacceptable product. To say that a process is stable does say that changing the output from that process will require a fundamental change to the process.

Construction

When Control Charts are established to understand the current operation of a process or to learn about the process so that future output from that process can be better than the past output, control limits (℄, LCL, and UCL) are <u>always</u> calculated from data collected from the process. Target values, specifications, and goals **do not** serve as limits for action. In addition, they do not provide guidance about the appropriate actions required to improve the process.

The formulas for calculating limits are deceptively simple. The arithmetic could be done by an elementary school student. The difficulty arises when you try to determine the appropriate observations to include in the calculations. Subgroups must be formed in such a way that the charts answer the questions that are being asked. There is a skill to identifying the appropriate method of subgrouping. This skill can be acquired through education and training more comprehensive than presented in this module. Be assured that the correct method of subgrouping will be determined by blending knowledge about the specific process with statistical knowledge (and will seldom result in the use of random samples from the end of a production line).

This means that every person does not need to be able to calculate control limits, but each person needs to recognize where charts are appropriate and understand what limits on these charts are saying about the process. This module will concentrate on recognizing the type of chart to use in specific situations and understanding how to use these charts to "describe" a process.

Control Charts can be divided into two major categories–charts for **attributes data** and charts for **variables data**. The most common charts in each of these areas will be summarized.

Charts for Attributes Data: The most common charts for attributes data deal with proportions or counts. There are two charts based on proportions ("p charts" and "np charts") and two charts based on counts ("c charts" and "u charts").

Proportions: Charts for proportions assume that each time a subgroup is collected each item in the subgroup can be classified in one of two categories (such as successfully meeting come criterion or not). Also, the classification of one item does not influence the classification of any other item. The number of items classified in one way is the measure of interest. If the total number of items in the subgroup and the number of items in the category of interest are known, the proportion of items in the category can be computed. (For example, if 6 out of 10 traffic lights are red when you reach them, you could say that 0.6 represents the proportion of lights that are red.) If the number of items in the category is plotted on the chart, an "np chart" is used. The "np chart" requires that the same number of items are included in each subgroup. If the proportion of items in the category is plotted on the chart, a "p chart" is used. The "p chart" can accommodate changing subgroup sizes.

Some applications for "p charts" and/or "np charts" include:

- proportion defective product produced on one machine (p chart)
- proportion "A's" recorded in a class (p chart)
- number of bottles (in a case) that leak (np chart)
- number of parking places empty at 8:00 a.m. (in a specific parking lot) (np chart)

Counts: Control Charts for counts assume that the number of occurrences of some phenomenon are being counted. In addition, these charts assume that when or where one occurrence took place does not influence when or where some other occurrence will take place. Most people will agree that the number of occurrences counted (in any situation) will depend on the quantity of material (or some other measure) examined. If the quantity of material examined remains the same from one point to the next, a "c chart" is used. If the quantity of material examined varies from one point to the next, a "u chart" is used.

Some applications for "c charts" or "u charts" include:

- the number of scratches on a surface (c chart if all surfaces checked are the same size)
- the number of times an instructor says "uh" in one minute (c chart) (a recent study found that science instructors say "uh" less than English instructors)
- the number of illustrations in a chapter (u chart since the length of a chapter is not constant–some chapters have more pages than others)

Charts for Variables Data: Control Charts for variables data are much more powerful than charts for attributes data. Since variables data involves some measurement of magnitude, the information provided can be more specific than anything provided by charts using attributes data. A "p chart" may indicate that 3% of your product is oversized. Control Charts based on measurements from the product may reveal that reducing the average by 0.01 inch would bring the product in line with expectations. Or, measurements from the product may indicate that items vary so much (from one item to the next) that the variability in the process must be reduced.

Dealing with measurable characteristics requires analyzing two characteristics of the data–the location and spread. Location refers to some average value that attempts to capture information about "the centering" of the values. By far, the most often quoted measure of location is $\overline{X}$. Spread refers to how values differ from each other. The Range (R) and standard deviation (s) are the most often quoted measures of spread. With small subgroup sizes there is little theoretical reason to select one of the measures of spread over the other. In process improvement studies (where small subgroups are the norm) the Range is the most often used–because it is so easy to understand.

Analyzing location and spread leads to the use of two Control Charts (together). When multiple observations are selected to form a subgroup, the traditional approach to analysis is through the use of $\overline{X}$ and R charts. When subgroups are appropriately chosen, the R chart answers questions about the spread in the observations within subgroups. The $\overline{X}$ chart answers questions about the location of the process.

In most "real world" settings obtaining measurements on every item produced by a process is impossible (because of time, cost, manpower, or the destructiveness of testing). Therefore, $\overline{X}$ and R charts attempt to use measurements from a portion of the output to determine how the underlying process is operating. Subgroups are selected time after time. In each subgroup, the spread in the observations is calculated and plotted on the R chart. Each one of these R's provides some measure of short term variability in the process. The concept of short term variability is shown in Figure 9.2.

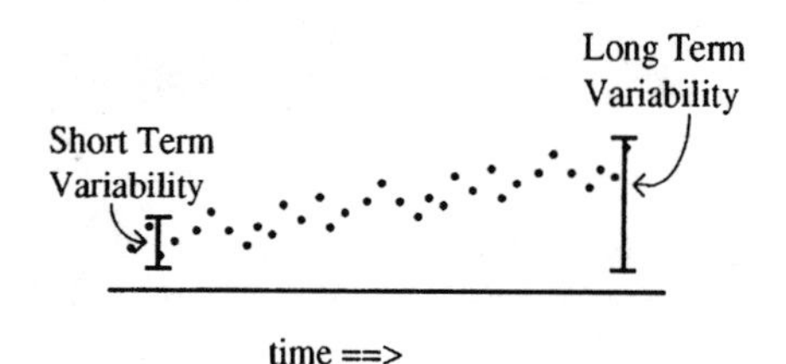

Figure 9.2 Short Term vs. Long Term Variability

The short term variability for this process would be approximately the same for any point in time (covered by this graph). With a measure of short term spread in the process, we can turn our attention to the question of where the spread is located as time passes. If the location does not change, the short term variation in the process will be the same as the long term variability. If the location is changing, as it is in Figure 9.2, variability in the $\overline{X}$'s should provide a signal. Therefore the R chart is asking if the short term variability is stable. When the R chart show stability, the $\overline{X}$ chart is asking if the long term variability is the same as the short term variability. If both of these conditions are true, the process is said to be **stable** or **in statistical control**. When the R chart does not show stability, calculation of limits on the $\overline{X}$ chart is not recommended.

Figures 9.3, 9.4, 9.5, and 9.6 provide more insight into the concept of process stability. For each diagram, consider what can be said about short term variability and long term variability.

In Figure 9.3 we see a process where the short term variability is remaining constant, and the average is not changing. As a consequence, the variability in the process (over the long term) is stable. The process is operating in a predictable fashion. If the conditions that produced this output can be expected to continue, the future output can be predicted. Predictability is one of the advantages of a stable process. Notice that the scale on the vertical axis is not needed to make this statement about stability. The measurements may range (consistently) from 0.00001 to 0.00002 or from 1000 to 2000. Stability implies that the range of observations can be forecast; not that the range is small or even desirable.

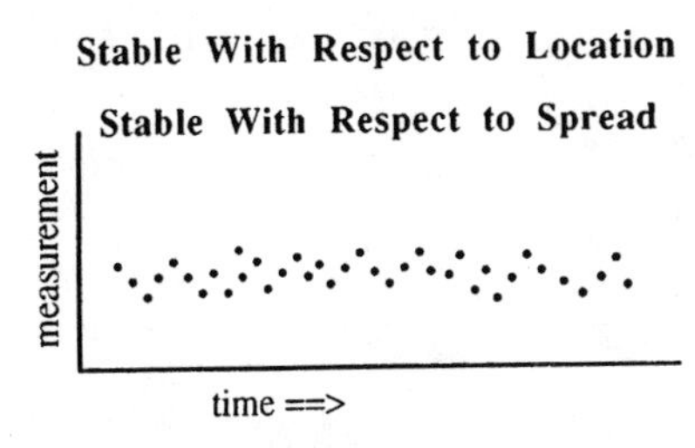

This process is **stable** or
in statistical control.

Figure 9.3 Variables Data–Stable Process

Figure 9.4 shows a process that is similar to the one shown in Figure 9.2. Again short term variability remains approximately the same, but the average changes. You would expect to see an R chart that shows control (stability) and an $\overline{X}$ chart that shows instability (lack of control). Prediction about the future output from this process is difficult without addressing the variability in the average.

Figure 9.5 shows a process with a stable average, but with inconsistent short term variability. The spread starts "small," becomes "larger," returns to "small," and explodes to larger than at any previous time. Data collected from this process should produce an R chart that signals instability. If a process is

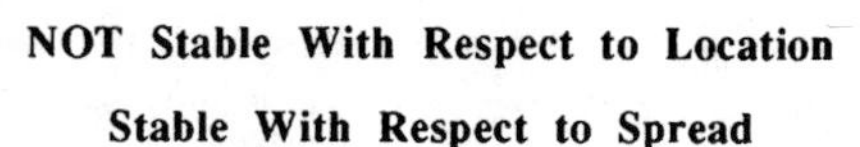

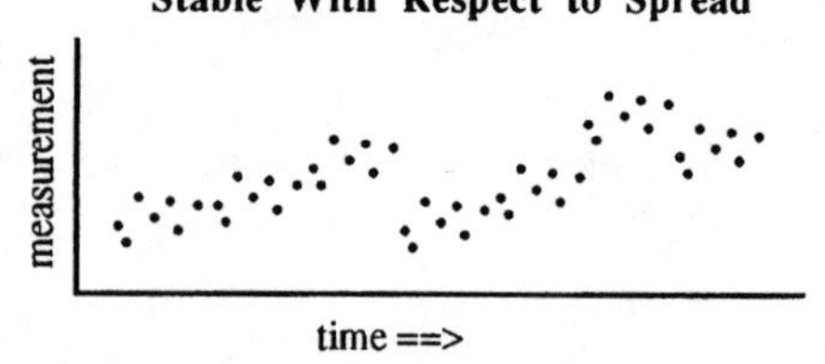

This process is **not stable,**
or it is **out of control.**

Figure 9.4 Variables Data–Not Stable (Location)

Stable With Respect to Location

NOT Stable With Respect to Spread

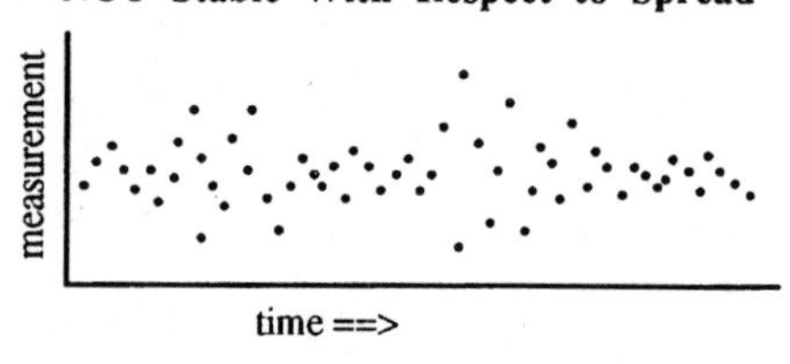

This process is **not stable,** or
it is **out of control.**

Figure 9.5 Variables Data–Not Stable (Spread)

stable, the Run Chart for the $\overline{X}$ values should not show any patterns. Predicting the future output from this process would be difficult.

The final possibility for a process is shown in Figure 9.6. Some people refer to this as a state of chaos. The reasons are fairly clear–the short term variation is changing over time and the average is changing. Attempts to predict future output from this process would be futile.

There are other charts for variables data in addition to $\overline{X}$ and R charts. When there is not a rational basis for forming subgroups, an X (or Individuals) chart is used along with a "moving range" chart. Individuals charts are most

NOT Stable With Respect to Location

NOT Stable With Respect to Spread

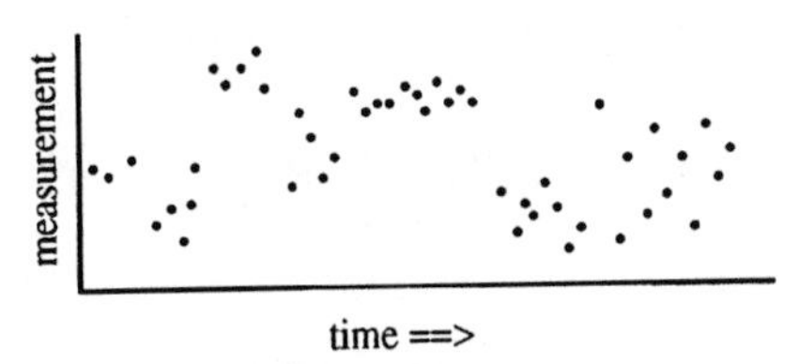

This process is **not stable,**
or it is **out of control.**

Figure 9.6 Variables Data–Not Stable (Chaos)

appropriate in cases where a single number is reported on some periodic basis. Some examples include:

- weekly sales
- quarterly profits
- expenses in some specific category
- grade point average
- time to drive to work

Summary: The situation will dictate the type of chart that is appropriate. When some characteristic can be measured (length, time, weight, etc.) rather than classified (good, bad) or counted more information about the process can be obtained. There is usually a trade off between the added information gained by measuring and the added difficulty in providing this measurement in a consistent manner. Once an approach is selected, each chart has its own set of formulas and considerations about subgrouping. Be sure to seek the assistance of someone with special training in this area.

Examples

Example 1 in Module 8 presented a Run Chart showing the number of scratches found when 30 desks were inspected each day for 20 days. To determine if the variability seen from day to day is within what should be expected from this process, control limits for a "c chart" should be calculated. These limits are shown in Figure 9.7.

Using the appropriate formulas for a c chart, the center line and control limits for this chart are:

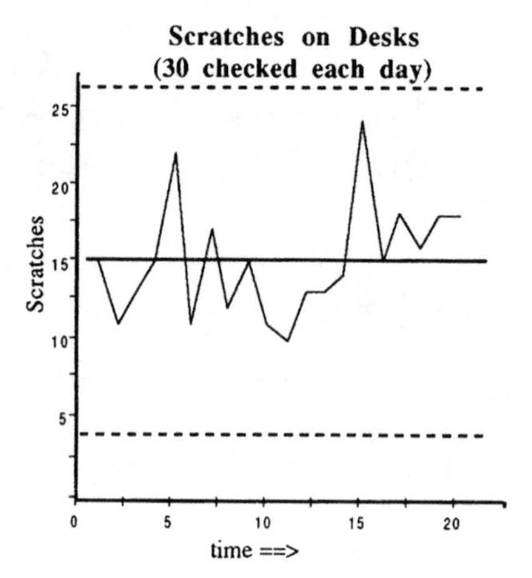

Figure 9.7 Sample "c chart"

LCL = 3.4
℄ = 15.0
UCL = 26.6

The c chart was selected because each point is obtained by counting the number of scratches on 30 desks. We are counting–indicating that this is attributes data. We are not dealing with proportions since we do not classify each desk as scratched or not; we count the number of scratches (recognizing that one desk may have 0, 1, 2, or more scratches). The number of desks remains constant–dictating the choice between the c and u charts.

Example 3 from Module 8 provides an example for using a Control Chart for variables data. In this example, monthly phone bills for one household were plotted. Since these are individual numbers that are reported on a periodic basis, the appropriate chart would be an Individuals (or X) chart. The chart is shown in Figure 9.8.

Using a moving range chart to estimate short term variability, the calculated center line and limits for the Individuals chart are:

LCL = 9.340
℄ = 26.963
UCL = 44.586

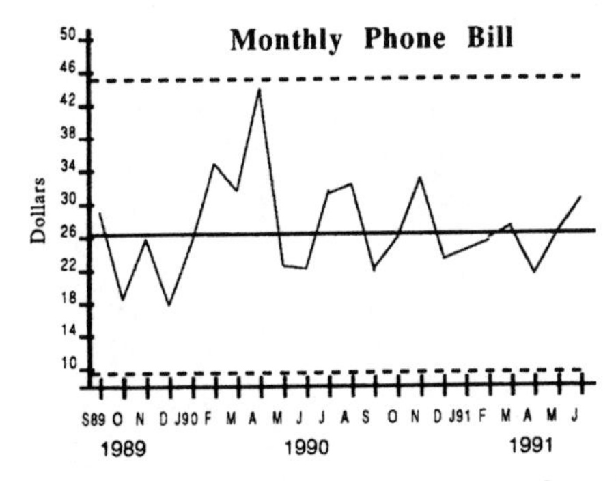

Figure 9.8 Sample "Individuals (or X) Chart"

Recall that we were tempted to look for a special cause for the high point for April 1990. The limits on this Control Chart indicate that this point should not be questioned as something special.

Evaluation

Control limits provide a method for assessing the amount of variability that should be expected in some characteristic. If a comparison between the value of the control limit and the value to be plotted provided all the information needed to assess process stability, there would be no need to draw the graph. Likewise, Run Charts provide a method for viewing patterns in data. If the Run Chart could determine when a plotted point was "too high" or "too low," there would be no need for control limits. Neither piece of information is complete without the other. Therefore, evaluation of Control Charts depends on looking at the Run Chart with the control limits added.

Processes that are stable will exhibit natural variation. The *AT&T Statistical Quality Control Handbook* uses the following description:

> These characteristics of a natural pattern can be summarized as follows:
>
> 1. Most of the points are near the solid centerline.
> 2. A few of the points spread out and approach the control limits.
> 3. None of the points (or at least only a very rare and occasional point) exceeds the control limits.

A natural pattern has all three of these characteristics simultaneously. The pattern will look unnatural if any one of the three is missing.

Patterns in the plotted points are referred to as "Runs," and the tests used to check for "runs" are called "runs tests." The tests attempt to operationalize the guidelines presented in the AT&T statement (what constitutes "most" or "a few?") The Runs Tests presented in the following material are a few of the more common tests used in industry. Hundreds more exist.

Test 1: None of the points should plot outside the control limits. If a point plots outside the control limits, effort should be directed to find the special cause. A Control Chart failing this test is shown in Figure 9.9.

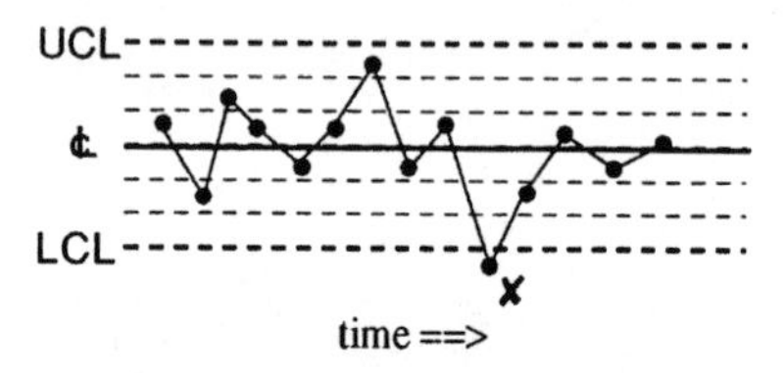

Figure 9.9 Point Beyond the Control Limits

Test 2: Most points should plot near the center line. (Or only a few points should plot near the control limits.) Consider the top half of the chart and the bottom half separately for this test. Divide the area between the center line and the control limit into three equal bands. If two out of three consecutive points plot in the outer most band, consider this as a signal to look for a special cause of variability. A Control Chart that fails this test is shown in Figure 9.10.

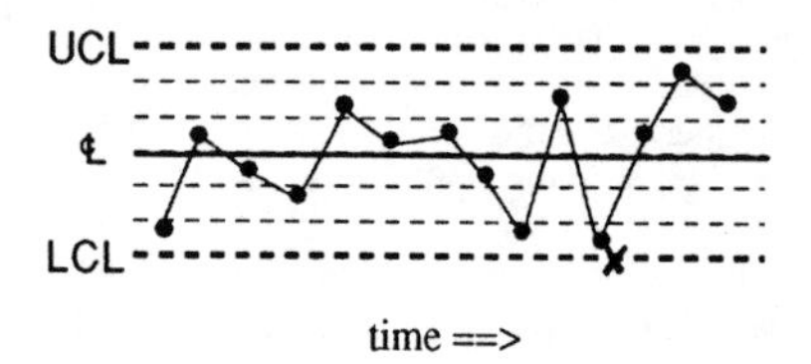

Figure 9.10 Two Out of Three Points "Near" The Same Control Limit

Test 3: Most (not all) points should plot near the center line. The points are said to be hugging the center line or showing less variability than should be expected if 15 points in a row plot within the closest bands (one above the center line and one below). This is not a signal of an exceptionally good process. This can be caused by people reporting results they think are expected (rather than what is actually occurring) or by the way subgroups are formed for $\overline{X}$ and R charts. Figure 9.11 shows a Control Chart that fails this test.

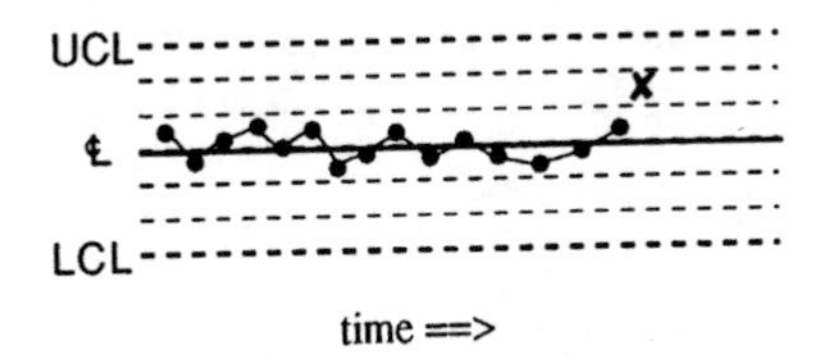

Figure 9.11 Hugging the Center Line

Test 4: About the same number of points should plot above and below the center line with random fluctuation. If more than 7 points in a row plot on the same side of the center line, a cause for the shift should be investigated. Figure 9.12 shows a Control Chart that fails this test.

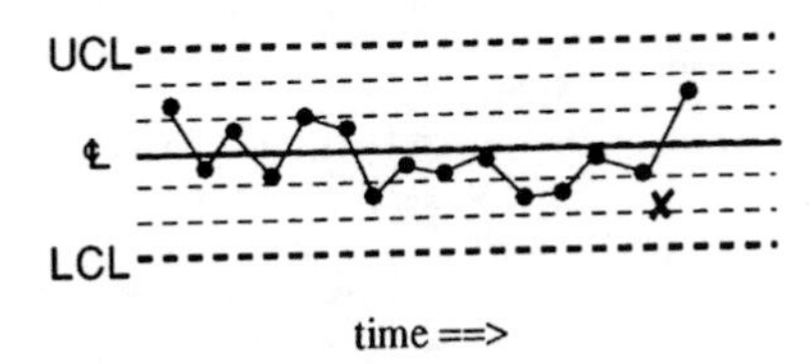

Figure 9.12 Sustained Shift

Test 5: Random fluctuation about the center line does not produce long upward or downward trends. Seven consecutive increases, or seven consecutive decreases, should be viewed as a signal that a trend exist. Notice that eight points are required to see seven increases. A trend is shown in Figure 9.13.

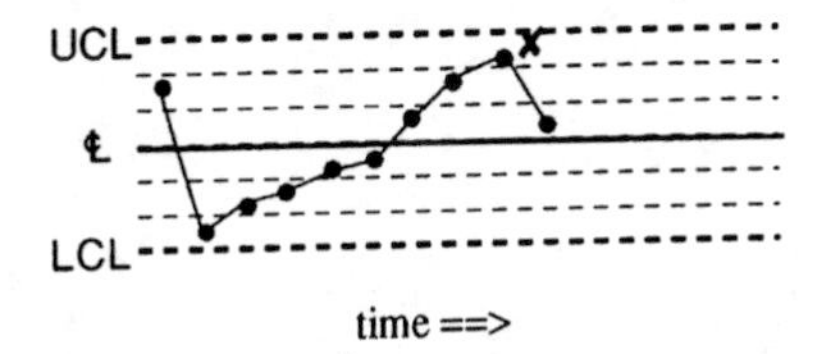

Figure 9.13 Trend

Test 6: No pattern should repeat itself in a predictable pattern. One common pattern is referred to as a "saw tooth pattern." This can occur when you attempt to plot two processes on the same chart. A saw tooth pattern is shown in Figure 9.14.

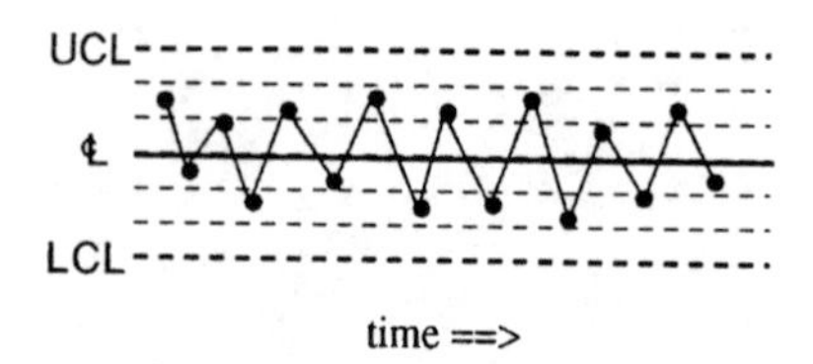

Figure 9.14 Saw Tooth Pattern

Cautions

Control limits for charts must be calculated from data obtained from the process if the information is to be used to provide direction for improvement. Specifications and targets **do not** belong on these charts–in fact they tend to encourage tampering and increased variability.

The effective use of Control Charts depends on the formation of subgroups. Much consideration should be given to the sources of variability present in a process and where this variability will show up on Control Charts. What sources will be hidden in the subgroups, what sources will show up between subgroups? The answers to these questions will aid in forming rational subgroups. These questions have not been addressed in this module. This is one reason that formulas for calculation of control limits have not been provided.

Control Charts tend to be one of the first process improvement tools used in industry. Many people jump into the formulas without collecting adequate knowledge of the process. Many sources of variability can be identified and eliminated through the development of Process Flow Diagrams and the development of operational definitions. Do not overlook the importance of these tools.

Next Steps

Control Charts provide guidance about the type of action required to improve a process. If a process is stable, a fundamental change in the day to day operation of the process will be required. These changes usually require some type of management action. Control Charts **do not** identify what change is required. Workers in the process being studied will usually have some suggestions about what actions to try. Control Charts can be used to evaluate the effect of these actions (simply continue plotting on the same Control Chart).

Exercises

NOTE: Formulas for calculating control limits have not been provided in this book. Until the issues of subgrouping (how to decide which items to include in the subgroup) are addressed, the simplicity of the arithmetic involved tends to provide a false sense of security. These issues are beyond the scope of this text.

9.1 Calculation of Control Limits for Exercise 1 in Module 8 results in the following:

UCL = 40.109

℄ = 35.588

LCL = 31.066

a. Determine the appropriate type of Control Chart to analyze this data.

b. Draw the Control Chart and perform the runs tests.

9.2 Calculation of Control Limits for Exercise 2 in Module 8 results in the following:

UCL = 0.23

℄ = 0.101

LCL = NONE

(A lower control limit listed as "NONE" means that it is not possible to observe a point on or below the lower control limit. "NONE" is reported when the calculated value is less than 0. Some people write this lower control limit as "0"–causing confusion when a value of 0 is observed.)

a. Determine the appropriate type of Control Chart to analyze this data.

b. Draw the Control Chart and perform the runs tests.

9.3 Calculation of Control Limits for Exercise 3 in Module 8 results in the following:

UCL = 6.3

℄ = 2.04

LCL = NONE

(A lower control limit listed as "NONE" means that it is not possible to observe a point on or below the lower control limit. "NONE" is reported when the calculated value is less than 0. Some people write this lower control limit as "0"–causing confusion when a value of 0 is observed.)

a. Determine the appropriate type of Control Chart to analyze this data.

b. Draw the Control Chart and perform the runs tests.

c. What assumption did you have to make to perform the runs tests (in order for them to be meaningful)?

9.4 Calculation of Control Limits for Exercise 4 in Module 8 results in the following:

UCL = 35.96

℄ = 29.75

LCL = 23.54

a. Determine the appropriate type of Control Chart to analyze this data.

b. Draw the Control Chart and perform the runs tests.

9.5 Calculation of Control Limits for Exercise 5 in Module 8 results in the following:

UCL = 35.22

℄ = 26.145

LCL = 17.07

a. Determine the appropriate type of Control Chart to analyze this data.

b. Draw the Control Chart and perform the runs tests.

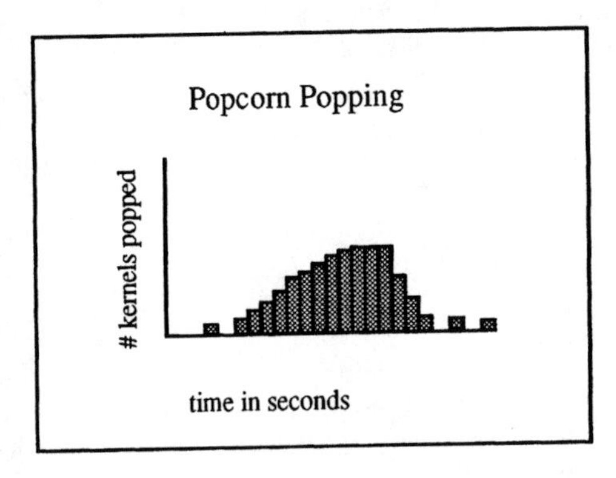

Module 10: Histograms

Histograms show the distribution of some quantitative measure of a characteristic. A Histogram will show how the observations "cluster." Since the order of production is lost when values of similar magnitude are grouped, Histograms (alone) do not provide a basis for describing the future output from a process. Histograms summarize the output of the process over the period of time that provided the data.

Construction

Histograms are used to describe a collection of quantitative data. Therefore, you should not be surprised to find them described in your textbook along with other descriptive tools. In this module, we will discuss the use of Histograms in process studies. Although the steps to construct a Histogram do not change, the eventual use of the chart must be considered. If the chart is going to be used to describe the expected future output from a process, the data used to construct the Histogram should be collected over time, plotted over time (Run Chart), and checked for stability (Control Charts). A stable process is necessary if the Histogram is going to be used for any predictions.

The method of constructing a Histogram varies from one individual to another. The goal is to display the data in enough detail to show the variability in the data without showing so much detail that the clustering of observations is lost.

Histograms usually follow a few conventions. All class widths are the same, every observation in the data set falls in one category, and the end points for each class are chosen so that there is no question which interval includes any particular observation. The following steps provide some "rules of thumb" that work in most cases.

STEP 0: Plot the data in time order. Check for trends or patterns. Do not use a Histogram to describe future output if trends or patterns are found.

MAKING THE HISTOGRAM

STEP 1: Determine the number of observations. Let "n" represent this quantity.

STEP 2: Determine the number of classes (or bars) that you will show. Your final chart may have slightly more or fewer classes than this because of rounding in later steps.

- If there are logical break points, use them. (For example: 90, 80, 70, 60, and below 60 provide logical breaks in reporting grades.) Skip to STEP 8.
- If logical breaks do not exist, the whole number closest to the square root of n usually provides a good estimate for the number of classes.

STEP 3: Determine the range for the entire data set. To do this, find the largest and smallest values. Then subtract the smallest number from the largest.

STEP 4: Determine the class width. To do this divide the range (found in STEP 3) by the number of classes (found in STEP 2). You may want to round this number up (slightly). The example in the next section of this module will illustrate this.

STEP 5: Determine the width of the Histogram that will result from your decisions in the previous steps. To do this, multiply the class width that you selected (after rounding in STEP 4) by the number of classes (in STEP 2). This width must be at least as large as the range.

STEP 6: Determine the lower end point for your Histogram. You would like to "center" your data. You can do this by finding the difference in the range (STEP 3) and the width of your Histogram (STEP 5). To find the lower end point of your Histogram subtract half of this difference from the smallest value in your data set.

STEP 7: Set up class intervals. The first class will start at the lower endpoint (determined in STEP 6). To find the upper endpoint add the class width (found in STEP 4) to the lower endpoint. The second class will begin <u>at the same point</u> that the first class ended. If the class breaks are values that could be observed, show the breaks to one more decimal place than the data.

STEP 8: Make a Check Sheet and place the observations in classes.

STEP 9: Set up the graph. Be sure to show a scale on both axes and label both axes. The vertical axis can represent the number of observations in a class (frequency) or the proportion of observations in a class (relative frequency). Give the chart a title.

STEP 10: Add bars to the graph. Depending on the scale on the vertical axis, the height of the bars will be determined by the number of observations in the class or the proportion of observations in the class.

These steps may sound complicated, but they are logical (once you see them used). Fortunately, commonly available computer software can help. Most spreadsheets will construct Histograms.

Example

The following data represent output from a process that produces wooden blanks that will be used to make rulers. Of course, the ideal length is 12.000 inches. There are 80 observations.

12.012	12.002	12.010	12.012
12.001	12.010	12.011	11.999
12.010	12.019	12.007	12.019
12.010	12.006	12.012	12.005
12.014	12.010	12.005	12.001
12.000	12.011	12.009	12.013
12.004	12.010	12.010	12.013
12.009	12.002	12.013	12.018
12.007	12.023	12.006	12.015
12.010	12.017	12.011	12.012
12.011	12.016	12.008	12.006
12.007	12.014	12.017	12.003
12.016	12.009	12.011	12.000
12.004	12.010	12.006	12.014
12.011	12.004	12.010	12.009
12.011	12.017	12.008	12.008
12.008	12.004	12.019	12.021
12.006	12.017	12.001	12.009
12.005	12.007	12.012	12.006
12.014	12.000	12.012	12.011

If this data is going to be used to predict future output from this process, the data should be plotted in the order of production. Since we are not given information about how the numbers (reported above) are ordered, further exploration would be needed to determine the time order of production. Figure 10.1 represents what we could find.

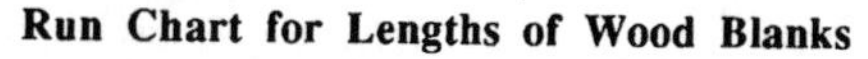

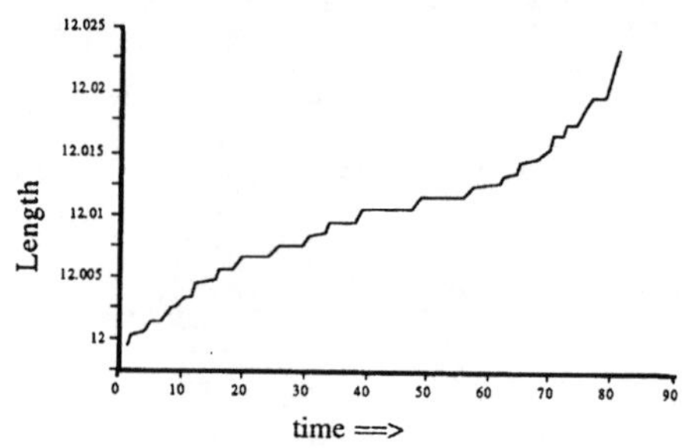

Figure 10.1 One Possible Run Chart

This Run Chart shows a clear upward trend. Predicting future output from this process would depend on our knowledge about the cause of this trend. Should be expect the trend to continue upward, or is the process allowed to "drift" for a certain length of time before it is "reset?" A Histogram based on this data would not represent what would be expected from this process in the near future.

Figure 10.2 represents another possible Run Chart that could result from this data. In this case, the 20 observations listed in column 1 came first in the order of production followed by the observations in column 2, then column 3, and finally column 4. This Run Chart does not show any trends or patterns. In addition, control limits would support a statement that this process appears to be stable. A Histogram will help us understand the output from this process.

STEP 1: There are 80 observations. $(n = 80)$

STEP 2: There are no logical breaks for this data. Therefore, use

$\sqrt{80} \approx 8.944$ Use 9 classes.

STEP 3: The largest value is 12.023.

The smallest value is 11.999.

The range is $12.023 - 11.999 = 0.024$.

Run Chart for Lengths of Wood Blanks

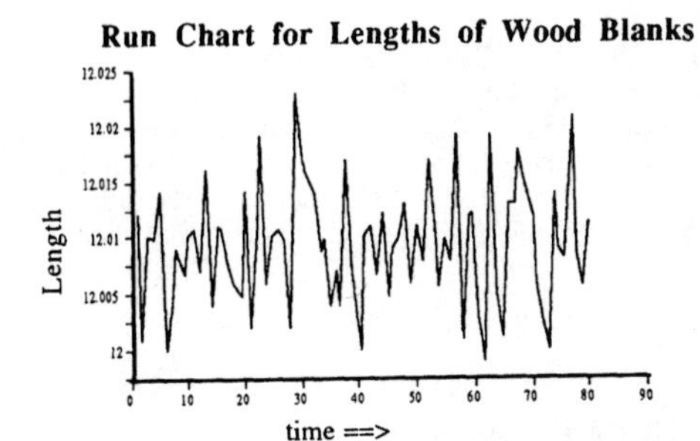

Figure 10.2 Another Possible Run Chart

STEP 4: $\frac{0.024}{9} \approx 0.00267$

Rounding this to 0.003 will provide a reasonable class width.

STEP 5: 0.003 x 9 = 0.027

STEP 6: The Histogram needs to cover a width of 0.024.

These calculations will lead to a Histogram with a width of 0.027.

(0.027 - 0.024)/2 = 0.0015

The lower end point of the Histogram should be at

11.999 – 0.0015 = 11.9975

STEP 7:
Class 1: 11.9975 – 12.0005
Class 2: 12.0005 – 12.0035
Class 3: 12.0035 – 12.0065
Class 4: 12.0065 – 12.0095
Class 5: 12.0095 – 12.0125
Class 6: 12.0125 – 12.0155
Class 7: 12.0155 – 12.0185
Class 8: 12.0185 – 12.0215
Class 9: 12.0215 – 12.0245

Since the data are reported to three decimal places, the class are breaks shown in the fourth decimal place. Each observation falls in one class, and the choice of class is clear cut.

STEP 8: Making the Check Sheet should involve one pass through the data. The first observation (12.012) is placed in a class, then the next observation (12.001) is placed in a class. This continues until all observations have been placed. The resulting Check Sheet is shown in Figure 10.3. **DO NOT** search the entire list for all observations in Class 1 before going on to Class 2.

Check Sheet for Lengths

Class	Tally
11.9975 - 12.0005	IIII
12.0005 - 12.0035	卌 II
12.0035 - 12.0065	卌 卌 III
12.0065 - 12.0095	卌 卌 IIII
12.0095 - 12.0125	卌 卌 卌 卌 III
12.0125 - 12.0155	卌 III
12.0155 - 12.0185	卌 I
12.0185 - 12.0215	IIII
12.0215 - 12.0245	I

Figure 10.3 Tabulation of Lengths

STEPS 9 and 10: If we choose to use the number of observations in each class on the vertical axis, the Check Sheet provides a quick way to find the heights of the bars. Class one should have a height of 4; class 2, a height of 7; class 3, a height of 13; and so on down the line to class 9 with a height of 1. The resulting graph is shown in Figure 10.4.

A quick glance reveals that most of the observations cluster around the middle with fewer and fewer observations as you move further from the center. This should not be surprising from a process of this type–although it is not guaranteed.

Evaluation

Histograms, even those constructed from processes that produce Run Charts that look stable, can assume a number of shapes. The shape of the Histogram can reveal some things about the process that produced the data. The following examples will illustrate.

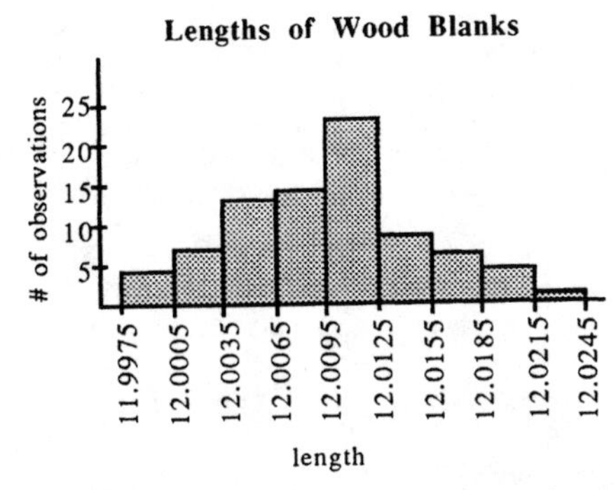

Figure 10.4 Histogram for Lengths

Most processes will produce Histograms with all observations clustered together in some fashion. The Histogram shown in Figure 10.5 could indicate that product has been sorted–the middle of the distribution appears to be missing.

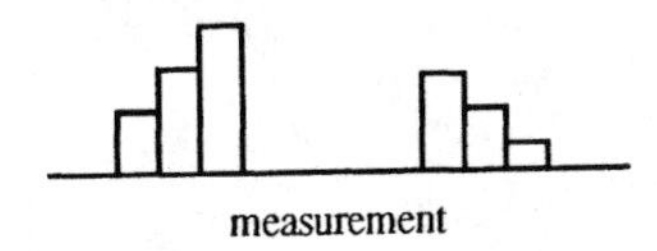

Figure 10.5 Histogram–Middle Missing

The numbers used to make a Histogram are the result of some measurement process. If the measurement device used by the data collectors does not show the level of precision requested, a Histogram similar to Figure 10.6 may result. For example, if a person is given a ruler marked in inches and asked to record the measurement to the nearest half inch, this stair step appearance may occur.

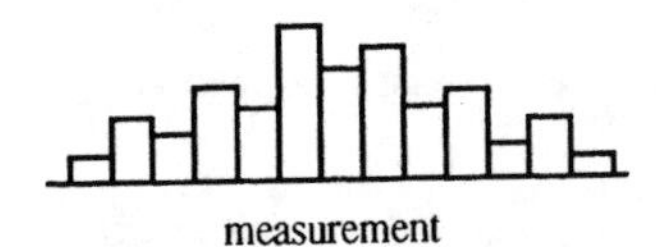

Figure 10.6 Histogram–Stair Steps

The choice of class width may be another possible cause for this pattern. Recall the example in Module 6 where we were collecting data about traffic flow through an intersection on the Virginia Commonwealth University campus. Suppose we looked at the number of pedestrians crossing in a thirty minute period. We might start recording at 7:45 a.m. on Monday morning. The first class interval would be 7:45 – 8:15; the second, 8:15 – 8:45; the third, 8:45 – 9:15; and so on. University classes are scheduled "on the hour" until "50 minutes past the hour." Naturally, more pedestrians would be observed during the intervals that include class changes.

Suppose you were making a Histogram of the proportion defective product seen in an inspection process that requires you to inspect 50 items each shift. Further suppose you chose the following classes:

Class 1: 0% to 2.5%
Class 2: 2.5% to 5.5%
Class 3: 5.5% to 8.5%
Class 4: 8.5% to 11.5%
• • •

All of the classes cover a 3% interval (except the first one where –.5% would not have any meaning). If you are inspecting 50 items and classifying each item as good or bad, the only possible outcomes from one subgroup would be 0%, 2%, 4%, 6%, etc. Notice how these "fall" into classes. Class 1 would include all observations of 0% and 2%; the only observations included in class 2 are 4%; class 3 would include all observations of 6% and 8%; only 10% would fall into class 4; and the pattern continues.

Another pattern that occurs is shown in Figure 10.7. Instead of trailing off on the sides, a "blip" appears. This can happen when there are "catch all" classes such as any measurement above 10 goes in this class. Another possible explanation is fear or denial of the existence of measurements outside a specific range. If these "blips" appear close to specification limits for some characteristic, there is reason for concern.

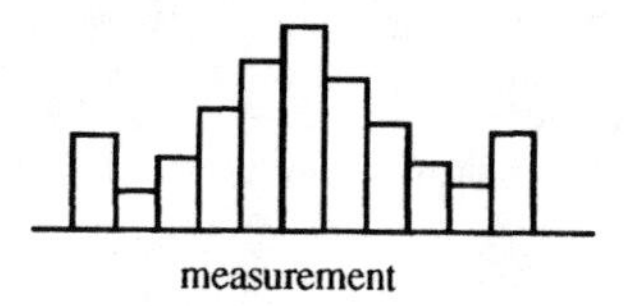

Figure 10.7 Histogram–Heavy Tails

If output from two different processes are combined, the resulting Histogram may resemble the one shown in Figure 10.8. Many times the combining of processes is not obvious. For example, two measurements are taken on the end of an aluminum pipe to determine the thickness of the pipe, or two of the same item made on different machines are measured, or two people are used to measure output from one machine.

Figure 10.8 Histogram–Two Modes

The Histogram shown in Figure 10.9 provides a less obvious pattern. This Histogram looks "flat." This can be caused by a process that is allowed to drift slowly between some limits. (Hopefully, this would have been caught with a Run Chart.) Constant (over) adjustment of a process can produce a similar Histogram. This type adjustment can occur when each individual observation (point on a Run Chart) is used as a basis for adjusting a process.

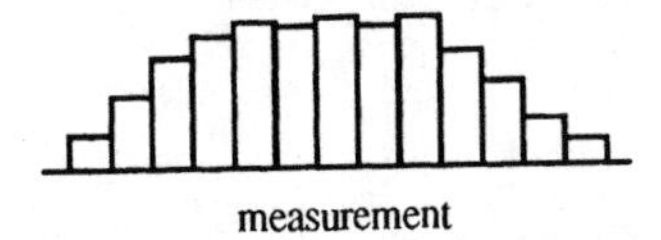

Figure 10.9 Histogram–Flattened

A final pattern is presented in Figure 10.10. This Histogram appears to be missing the left side and stretched to the right. In statistical jargon, it is skewed to the right or positively skewed. In some situations a natural boundary prevents measurements outside some interval (for example, the diameter of a hole cannot be less than 0 inches). This would account for the missing left tail. But if this were the case, you would probably expect to see more observations in the first class. The long right tail could be caused by a process that has started to drift, or it may be a natural occurrence for the process. For example, a Histogram showing life expectancy for many products (such as light bulbs) will usually appear to be "stretched."

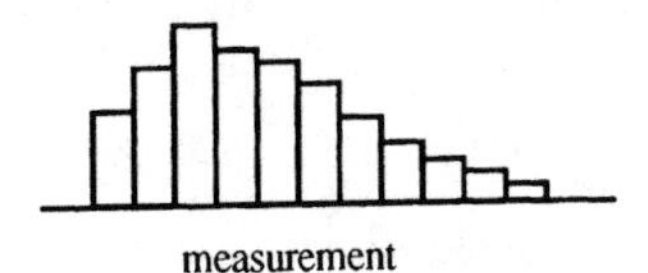

Figure 10.10 Histogram–Skewed

These are only a few of the possible patterns that can be seen in Histograms. In fact, each Histogram will have its own unique shape. If a different number of classes is selected or different class breaks are used, the same data can produce two Histograms that look different.

Cautions

Histograms summarize data without respect to the time order that produced the observations. The shape of the Histogram **does not** provide evidence of process stability. Conversely, process stability does not guarantee any specific shape for a Histogram.

Next Steps

Run Charts, Control Charts, and Histograms can be used to describe output from a process or to study variability in the input to a process. Real improvement in processes occurs when the relationships between the inputs and outputs are known. Methods that allow us to study more than one variable at a time will help. One tool that aids in this study is the Scatter Diagram.

Exercises

10.1 Some insurance companies maintain telephone lines for agents to obtain quotes. Calls come into this line throughout the day. One operator was observed to see how much time was required to provide a quote using the current system. The times observed are reported below. (Observations are listed in columns and reported in seconds.)

113	110	86	122
83	99	107	117
107	124	131	105
107	91	95	106
118	108	106	90
79	109	109	98
89	101	106	89
102	95	84	125
98	92	141	99
107	118	126	79

a. Plot the data in a Run Chart.

b. If appropriate, draw a Histogram.

10.2 All mail coming into a large company enters through the Mail Room. Since the number of pieces of mail is some indicator of the amount of business generated by the company, the Mail Room counts the number of pieces of mail arriving. The following data were collected over a 10 week period.

Week 1	
M	498
T	267
W	320
R	316
F	304

Week 2	
M	484
T	311
W	308
R	306
F	316

Week 3	
M	494
T	295
W	298
R	279
F	294

Week 4	
M	486
T	312
W	308
R	310
F	314

Week 5	
M	530
T	330
W	310
R	290
F	291

Week 6	
M	496
T	312
W	300
R	316
F	293

Week 7	
M	520
T	316
W	283
R	288
F	295

Week 8	
M	480
T	298
W	311
R	297
F	301

Week 9	
M	483
T	307
W	292
R	291
F	322

Week 10	
M	465
T	300
W	296
R	304
F	308

a. Plot the data on a Run Chart.

b. If appropriate, draw a Histogram.

10.3 The in town miles per gallon have been recorded for a car. The results are shown (in columns) on the next page.

10.3 Continued

31.4	32.0	29.8	32.1
30.2	31.4	31.1	30.1
27.1	30.7	31.8	32.2
34.8	30.6	30.4	30.8
28.3	32.1	32.9	31.7
29.2	27.3	27.7	31.3
30.3	27.2	29.0	27.6
29.4	28.3	30.1	28.8
31.6	29.7	29.9	32.3
30.5	27.8	31.7	30.9
31.2	34.3	29.1	29.2
31.5	31.2	27.9	31.5
31.9	31.7	30.5	28.4

a. Plot the data on a Run Chart.

b. If appropriate, draw a Histogram.

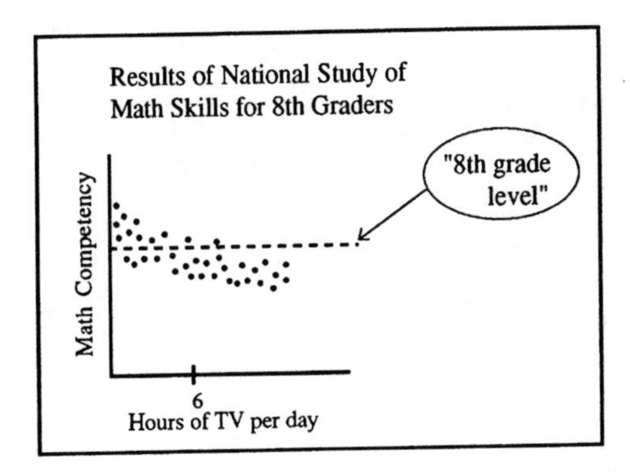

Module 11: Scatter Diagrams

Scatter Diagrams are used to graphically display the relationship between two quantitative variables. Scatter Diagrams are often used to direct model building efforts in regression analysis. These same diagrams can be used as descriptive tools. Scatter Diagrams do not prove cause and effect relationships–they can provide support for hypothesized relationships.

Construction

This module will concentrate on the use of Scatter Diagrams in process studies. Scatter Diagrams will allow us to see relationships between inputs and outputs. Before data can be collected a relationship is hypothesized, an input variable is identified, an output variable is identified, and a method of measuring both variables is developed. Data for these diagrams are collected in pairs. Each ordered pair provides the coordinates for one point on the diagram. Scatter Diagrams, like Histograms, do not show the time order of the data–therefore, process stability is assumed.

The horizontal axis (the "x axis") on the graph represents the input variable. You want to know how the output from the process changes as you change this input variable. This will help determine the "best" way to operate this process. The vertical axis (the "y axis") on the graph represents the output variable–the variable that you would like to be able to predict. You would like to know how the value of the output variable will change for a small change in the value of the input variable; or conversely, you would like to know that changing the value of the output variable by some desired amount can be accomplished by changing the value of the input variable by a known amount.

Example

The managers of one large company hypothesized that the number of accidents reported in their plant increased as the average amount of overtime increased. That is, they felt that when people worked overtime they became tired, and tired people were more prone to accidents. They decided to collect data and plot a Scatter Diagram to start to evaluate their hypothesis. The following data was collected over a period of 38 weeks when overtime was used.

OT	Accidents	OT	Accidents
6.1	12	4.7	9
4.9	12	6.4	10
6.2	9	6.2	8
7.2	11	7.4	13
7.2	12	7.0	12
7.6	13	8.3	14
6.5	13	4.3	9
6.2	10	6.0	11
7.0	13	6.6	12
7.3	10	8.1	12
4.5	10	7.9	12
8.2	16	5.9	12
5.3	10	7.0	12
11.9	18	5.7	11
8.6	15	5.5	10
9.8	17	7.0	12
9.3	14	8.9	16
8.5	13	8.5	14
6.7	11	5.4	9

Since the managers believed that they could predict accidents it they knew how much overtime was used, "overtime (OT)" was shown on the horizontal axis. "Number of Accidents" were recorded on the vertical axis. The scale on the overtime axis was shown in hours and the scale on the accidents axis was shown as a count. The first ordered pair showed an average of 6.1 hours of overtime and 12 accidents. This corresponds to a point 12 units above 6.1. Each successive ordered pair provides coordinates for another point. Notice that 7.0 hours of overtime and 12 accidents were seen on three separate occasions. Each time a point needs to be duplicated, a circle is drawn around the original point. The resulting Scatter Diagram is shown in Figure 11.1.

The Scatter Diagram shows a fairly strong relationship between these two variables. It does not prove that increasing overtime will increase the number of accidents. There could be some other variable that affects both of these

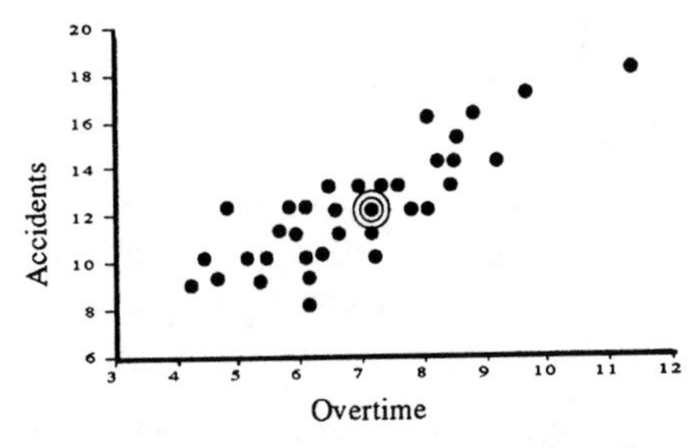

Figure 11.1 Accidents vs. Overtime

variables. In this example the company may only experience overtime when a particularly difficult product is in demand. Increase in demand causes more overtime. Increase in demand also causes more difficult work which results in more accidents.

Evaluation

Scatter Diagrams can be used to show the strength or type of relationship between two variables. Figure 11.1 showed a linear relationship. Graphically, if the points cluster in a way that resembles a straight line, we say there is a linear relationship. Not all relationships are linear.

Electric companies must be able to forecast demand for electricity. One variable that is believed to be an indicator of the amount of electricity that will be demanded is the forecasted high temperature for the day. During cold weather (winter), people use their heater. As the days warm up (spring), people turn off their heat and open the windows. When hot weather arrives (summer), demand for power goes up as people turn on the air conditioners. We expect to see a curve on the Scatter Diagram. An equation describing this curve would include a quadratic (x^2) term. This relationship is shown in Figure 11.2.

Figure 11.3 shows a different type relationship. In this example, we see the effect that increasing the speed of a machine (rpm's) has on a critical measure of some part. In this case, increasing the speed does not change the average measurement for the characteristic, but it does appear to introduce increased variability into the measurements.

The Relationship Between Daily High Temperature and Demand for Electricity

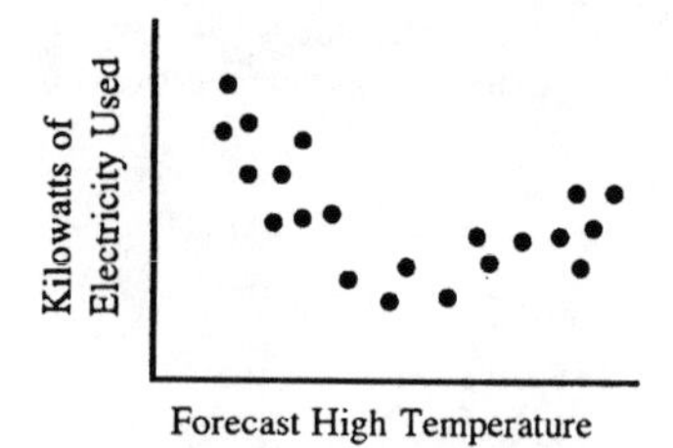

Figure 11.2 Kilowatts vs. Temperature

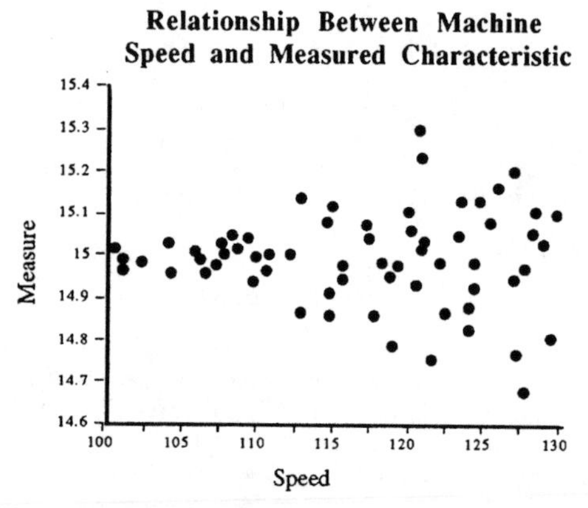

Figure 11.3 Measure vs. Speed

The next series of Scatter Diagrams illustrates how a third variable can be introduced into the analysis. Figure 11.4 shows an initial Scatter Diagram illustrating the relationship between the productivity of a plant and the number of hours that have passed since the shift started work. From this view productivity appears to remain approximately the same throughout the shift with large variability from day to day. To see this, pick any point on the horizontal axis. Then look at the spread in the points directly above that point. You may say that there is more variability at the beginning and end of the shift than in the middle.

Managers and workers questioned these results. This was not what they expected to see. Some claimed that people tired at the end of a shift, others

Productivity as Shift Progresses

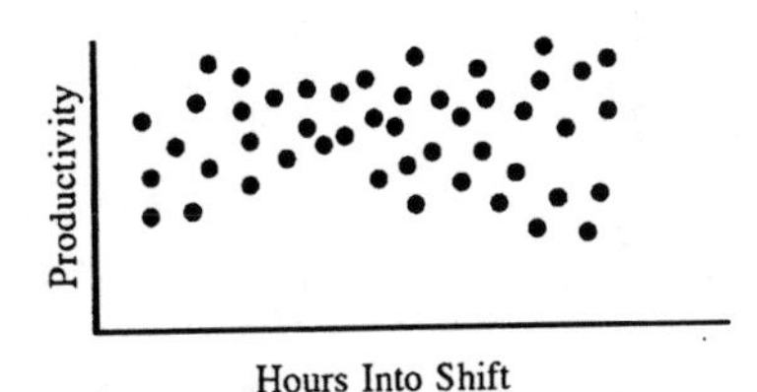

Figure 11.4 Productivity vs. Hours into Shift

claimed that it took a while for people to "get going" at the beginning of a shift. As more and more people commented on the results a pattern started to emerge. Workers on the day shift talked about coming in early in the morning "still asleep." Workers from the night shift talked about difficulty getting their "body clock" set to work between 3:00 a.m. and 7:00 a.m. The Scatter Diagram was revised to use different symbols for the different shifts. The result can be seen in Figure 11.5.

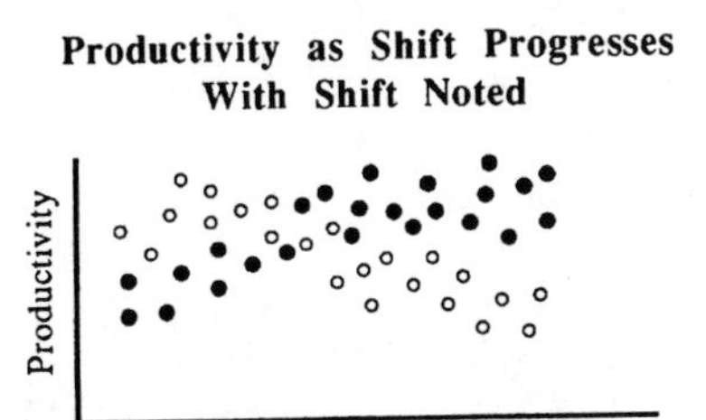

Figure 11.5 Productivity within Shift

The pattern is a little cloudy, but there does appear to be different patterns in the points produced by the different shifts. Separate Scatter Diagrams would make the analysis clearer. Figure 11.6 shows how productivity changes over the day shift. Figure 11.7 show the same information for night shift.

Clearly, the productivity on day shift increases through the first part of the shift. Productivity appears to remain fairly constant over the last half of the shift. This does not say that the workers work harder later in the shift. We are

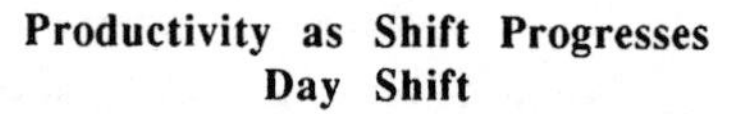

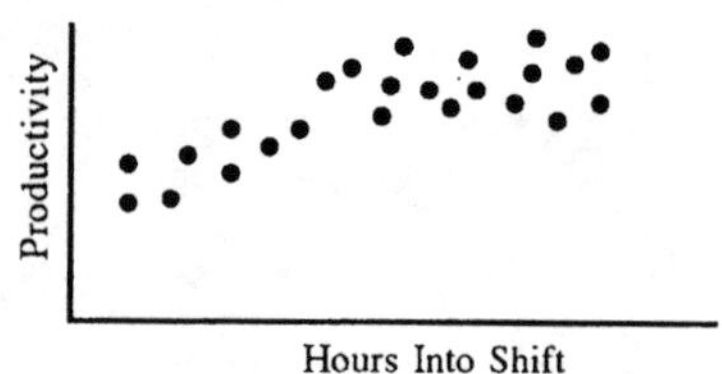

Figure 11.6 Productivity for Day Shift

not told how productivity is calculated, how jobs are scheduled, or how training and maintenance or factored into this measure.

Productivity on night shift appears to follow a different pattern. Information obtained from workers in this process may confirm that fatigue is a factor. Or, just as with the day shift, the way productivity is measured may produce this pattern.

Productivity as Shift Progresses
Night Shift

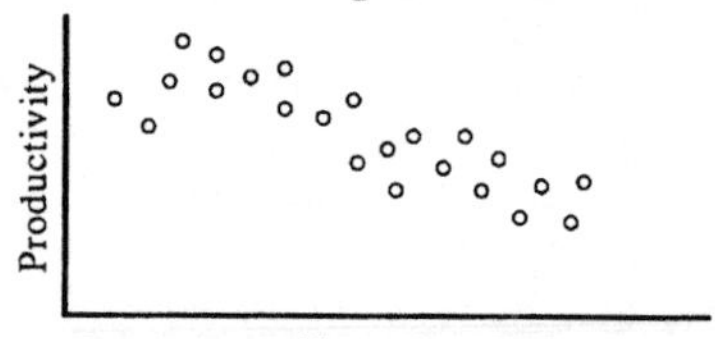

Figure 11.7 Productivity for Night Shift

Cautions

Cause and effect relationships can be hypothesized by people with knowledge about the process being studied. Data can be collected to provide support or to dispute their claim. Repeated data providing similar information may strengthen an argument. Scatter Diagrams document the strength of a relationship and show how variables tend to "move together," but causation **cannot** be proved with a Scatter Diagram.

Process stability, with respect to the input variable and the output variable, is assumed. Run Charts and Control Charts should be constructed and checked for patterns before the Scatter Diagram is constructed.

Next Steps

When relationships between changes in inputs and changes in outputs can be identified, teams can start using this information for designing the operation of the process. As knowledge is refined the critical inputs can be controlled leading to predictable improvement in the output from the process. Making improvements in early stages of the process will lead to higher quality, less rework, and lower costs. Integrating the use of the tools presented in these modules into the PDSA Cycle (Module 3) will help people work more objectively. The result of this objectivity should be increased teamwork and a better working environment.

Exercises

11.1 Companies that depend on orders placed over the phone try determine some measures that will predict sales. One company believed that the length of a call would be some indicator of the sales generated by the call. Use the following data to draw a Scatter Diagram showing the relationship between length of call (in seconds) and income from the call. Comment on what you see.

Length (in seconds)	Sales (in dollars)	Length (in seconds)	Sales (in dollars)
140	323.08	155	320.25
128	325.88	174	344.93
157	321.00	166	345.65
149	330.76	159	330.39
151	317.56	158	321.00
111	301.66	164	354.59
168	351.91	167	331.06
174	328.20	144	326.80
175	322.69	157	307.09
145	332.30	158	352.09
103	294.81	127	282.18
162	334.01		

11.2 The time required to complete an in class test varies from one student to another. One teacher (for a statistics class) believes that the time required to complete a test is related to the number of homework problems completed by the student. Use the data below to construct a Scatter Diagram. Comment on what you see.

# HW Problems	Time (in minutes)	# HW Problems	Time (in minutes)
42	48	13	70
30	62	36	59
11	72	2	77
50	50	20	71
41	53	37	54
15	71	10	72
21	66	42	54
35	57	33	59
35	60	20	68
11	72	5	75
18	69	18	70
8	73	8	74
49	49		

11.3 Recently, a large printing company has received a number of complaints about over billing (i.e., the customer says the bill is higher than it is suppose to be). The company management believes that the quotes for printing may be the source of the problem. The data presented below represent the quotes on the most recent 20 printing jobs and the actual bills for those jobs. Draw a Scatter Diagram showing the relationship. Comment on the process.

Quote	Bill	Quote	Bill
1697.30	2059.94	1449.27	1794.60
1498.90	1846.75	1823.57	2341.19
1479.59	1949.52	1599.33	1958.88
1558.29	1855.49	1614.38	1787.70
1377.34	1520.83	1570.01	1848.78
1518.49	1966.01	1247.87	1390.08
1413.39	1753.51	1466.07	1700.22
1908.89	2289.79	1634.64	2193.25
1758.11	2038.79	1620.70	2068.16
1492.00	1652.01	1750.71	2318.63

11.4 A convenience store operator wants to establish a way to predict hot chocolate sales for the store. An employee suggests using the forecasted low temperature as a predictor for the number of cups of hot chocolate that will be sold. Use the data below to develop a Scatter Diagram showing the relationship. Comment on what you see (Be careful).

Temp.	Hot Cocoa	Temp.	Hot Cocoa
37	35	36	31
45	47	50	48
49	47	46	45
38	40	44	43
34	34	49	44
39	35	38	38
33	27	36	36
38	32	44	41
38	33	47	49
31	31	32	28
40	39	42	39
45	45	46	43
43	42	40	39

If the statistics are boring, then you've got the wrong numbers. Finding the right numbers requires as much specialized skill–statistical skill–and hard work as creating a beautiful design or covering a complex news story.

Edward R. Tufte

Module 12: Data Collection

All of the tools presented in the previous modules assumed that the necessary data was available and meaningful. Historical data may exist, but if you do not know how it was collected and what conditions existed when the data was collected, be suspect–a better name for this type of data may be "hysterical" data. Obtaining good data requires time, effort, and careful planning.

Data collection should be viewed as a process. Many inputs are required to successfully collect data. The data collected (the output from the process) become the input to some data analysis process. Sophisticated statistical approaches to analyzing data cannot correct problems introduced in the data collection stage. On the other hand, careful planning and execution of the data collection step can make data analysis more straightforward.

Data collection can be summarized as a five step process[1]. Scholtes outlines these steps as

1. Clarify data collection goals
2. Develop operational definition
3. Plan for data consistency and stability
4. Begin data collection
5. Continue improving measurement systems

A careful analysis of these steps will reveal a heavy emphasis on understanding the measurement process that is being used. Whenever measurements are combined in any manner, we must ask the age-old question, "Am I combining apples and oranges?"

Another approach to planning for data collection comes from the news reporting area. Reporters guidelines remind them to include the who, what, when, where, why, and how of stories. In addition to considering these questions, a good data collection plan will consider what could go wrong, reduce the likelihood of these things happening, and determine how to react if these things do happen.

Each question leads to additional questions. The following check list (of questions) should help identify some of the areas that must be considered before the first measurement is obtained. The "why" questions should come first. The order that the other questions are considered will vary.

Why?

- Why am I collecting the data? (If no one can answer this question, do not continue. You are wasting time, effort, and money.)
- What am I (or the person requesting the data) trying to accomplish?
- Who will benefit from the study that uses the data collected?

What?

- What results do I expect to see from this study? (What relationships do you hypothesize before the study begins?)
- What am I trying to measure? (Concept)
- What variables can be used to measure this concept?
- What operational definitions will be used?
- What data are needed to make a decision?
- What level of measurement is required? (Is attributes data acceptable, or are measurements needed?)
- What form will be used to record data?
- What other information should be recorded–shift, date, time, worker, machine, person taking the measurement, etc.?
- If I had the data I just identified, how would I (or someone) analyze it? Have I identified the "right" data to collect?

Who?

- Who will design the forms for collecting data?
- Who will determine what to measure, when and where to measure, and how to measure?
- Who will collect the data?
- Who will analyze the data?
- Who will react to the study conducted using this data?
- Do all of these people have the same understanding about why the study is being conducted?
- Are all of these people using the same operational definitions?
- Who needs to be aware that this data is being collected?
- Who needs to approve this study? How do I get their approval?

How?

- How will I get the data?
- Physically, how will the collection process be done?
- How will people be trained to participate?
- How will I make sure that everyone collecting data is collecting data in the same way?
- How do I know if the measurement device(s) provide consist results? If one person measured the same item twice with the same measurement device, would they get the same result?
- How are "items" selected for measurement?
- How many measurements are needed?
- How will the data collection process "interrupt" regular operations?

When and Where?

- When will the data be collected? At one point in time or over some period of time?
- Where will the data be collected? At a supplier's facility, in-house, at a customer's location?
- When will measurements take place? At what point in the process?
- Does the product or process change as a result of data collection? (the psychology of knowing that something is being observed)
- Does the characteristic being measured change depending on the time it is measured? (due to shrinking, expanding, or destruction)

What If?

- What could go wrong?
- How can I minimize the occurrence of these events?
- How can I minimize the effect of something going wrong?
- How will discrepancies be reported?

These questions are not all inclusive. Every situation will have its own special points. Consideration of these points and the specific situation before the actual collection of data starts should simplify efforts in the collection and analysis steps of a decision making process.

Many plans for data collection look good on paper, sound good to the people designing the collection process, and flop when attempted! (The IRS W-4 provides a famous example.) Before undertaking a massive data collection process, a small scale (pilot study) should be conducted. The results of the study should be used to improve the data collection process. After the large scale collection process begins, continued effort is needed to insure that the plan is being followed.

Summary

Data analysis provides a rational basis for planning. The choice of analysis technique and the usefulness of the results obtained from data analysis are dependent on the data collection process. It is impossible to place enough emphasis on the importance of collecting meaningful data. Increasing efforts in planning for data collection and insuring consistent measurement procedures are easy to justify (in the end). Without meaningful data, the tools presented in the other modules lose their meaning.

Exercises

12.1 Suppose you are the manager of one of the following fast food restaurants: Arby's, Burger King, Hardee's, McDonald's, or Wendy's. You recognize that a large portion of your sales come from the "Drive Thru." Design a data collection plan to study "service" at the Drive Thru. What data would need to be collected? Be sure to operationally define all terms. Address the who, what, when, where, why, and how questions suggested in this module. *NOTE: Your assignment is **not** to collect data. Your assignment is to develop a plan for collecting data.*

12.2 Suppose that you work for a group concerned with highway safety. This group is concerned about the speed of cars on the highway. Develop a data collection plan to analyze this. Be sure to address the who, what, when, where, why, and how questions. *NOTE: Your assignment is **not** to collect data. Your assignment is to develop a plan for collecting data.*

12.3 Develop a data collection plan for studying the temperature in the building. Be sure to address the who, what, when, where, why, and how questions. *NOTE: Your assignment is **not** to collect data. Your assignment is to develop a plan for collecting data.*

12.4 Develop a data collection plan for studying the types of beverages served on the university campus. Be sure to address the who, what, when, where, why, and how questions. *NOTE: Your assignment is **not** to collect data. Your assignment is to develop a plan for collecting data.*

---

1 Scholtes, Peter R., *The Team Handbook: How to Use Teams to Improve Quality*, Joiner Associates, Inc., 1988.

Index